Springer Tracts in Nature-Inspired Computing

The book series is aimed at providing an exchange platform for researchers to summarize the latest research and developments related to nature-inspired computing in the most general sense. It includes analysis of nature-inspired algorithms and techniques, inspiration from natural and biological systems, computational mechanisms and models that imitate them in various fields, and the applications to solve real-world problems in different disciplines. The book series addresses the most recent innovations and developments in nature-inspired computation, algorithms, models and methods, implementation, tools, architectures, frameworks, structures, applications associated with bio-inspired methodologies and other relevant areas.

The book series covers the topics and fields of Nature-Inspired Computing, Bio-inspired Methods, Swarm Intelligence, Computational Intelligence, Evolutionary Computation, Nature-Inspired Algorithms, Neural Computing, Data Mining, Artificial Intelligence, Machine Learning, Theoretical Foundations and Analysis, and Multi-Agent Systems. In addition, case studies, implementation of methods and algorithms as well as applications in a diverse range of areas such as Bioinformatics, Big Data, Computer Science, Signal and Image Processing, Computer Vision, Biomedical and Health Science, Business Planning, Vehicle Routing and others are also an important part of this book series.

The series publishes monographs, edited volumes and selected proceedings.

Sanjay Chakraborty · Lopamudra Dey

Multi-objective, Multi-class and Multi-label Data Classification with Class Imbalance

Theory and Practices

Springer

Sanjay Chakraborty
Department of Computer and Information Science (IDA), REAL, AIICS
Linköping University
Linköping, Sweden

Department of Computer Science and Engineering
Techno International New Town
Kolkata, West Bengal, India

Lopamudra Dey
Department of Computer Science and Engineering
Meghnad Saha Institute of Technology
Kolkata, West Bengal, India

Department of Biomedical and Clinical Sciences
Linköping University
Linköping, Sweden

ISSN 2524-552X ISSN 2524-5538 (electronic)
Springer Tracts in Nature-Inspired Computing
ISBN 978-981-97-9621-2 ISBN 978-981-97-9622-9 (eBook)
https://doi.org/10.1007/978-981-97-9622-9

This Springer imprint is published by the registered company Springer Nature Singapore Pte Ltd.
The registered company address is: 152 Beach Road, #21-01/04 Gateway East, Singapore 189721, Singapore

To our Parents, Sister, and our Son Arohan
for their love and inspiration.

—Dr. Sanjay Chakraborty

—Dr. Lopamudra Dey

Preface

In the ever-evolving landscape of machine learning, the challenges of multi-objective, multi-class, and multi-label data classification are becoming increasingly critical, especially in the presence of data imbalance. This book delves into the intricacies of these advanced classification problems, providing a comprehensive exploration of methodologies, algorithms, and real-world applications. By addressing the unique difficulties posed by data imbalance, we offer practical solutions and innovative approaches to enhance model performance and accuracy. Whether you are a researcher, data scientist, or practitioner, this book aims to equip you with the knowledge and tools necessary to navigate and excel in the complex domain of multi-objective and multi-label classification, ultimately fostering a deeper understanding and more effective handling of diverse and imbalanced datasets. This book is divided into six chapters.

Chapter 1 introduces the fundamentals of classification in machine learning, beginning with an overview of the basics and the various challenges posed by class imbalance. Then four key classification approaches: binary, multiclass, multi-label, and multi-objective are explored. To provide a solid foundation, we delve into popular classification algorithms including support vector machines (SVM), decision trees, Naive Bayes, and neural networks, particularly focusing on their application in multi-class classification scenarios. This chapter aims to equip readers with a comprehensive understanding of classification techniques and their practical implementations in handling diverse classification problems.

Chapter 2 delves into the critical challenges of data irregularities and class imbalance in machine learning classification, exploring issues such as missing, incomplete, inconsistent, and noisy features, and discussing strategies for feature normalization. It addresses the pervasive problems of overfitting and underfitting, highlighting their impact on model performance. Furthermore, the chapter examines how class imbalance and data irregularities affect multiclass, multi-objective, and multi-label classification, offering practical solutions to mitigate these issues and improve the robustness and accuracy of classification models.

Chapter 3 provides a comprehensive overview of multiclass classification, beginning with the basics of binary classification and expanding into the nuances of multiclass classification, highlighting their pitfalls and diverse applications. We delve into the two main types of multiclass classification approaches—one-vs-one and one-vs-all—discussing their benefits and limitations through real-world case studies. The chapter also addresses the challenges of performing multiclass classification with imbalanced datasets, offering practical insights and hands-on coding examples to equip readers with the skills needed to implement effective multiclass classification solutions in real-world scenarios.

Chapter 4 delves into the complexities of multi-label and multi-objective classification, beginning with a discussion on the application of multi-objective SVM for multiclass classification. It explores advanced techniques for multi-objective and multi-label feature selection, providing a thorough understanding of how to optimize feature sets for complex classification tasks. The chapter also covers multi-label classification in image analysis and multi-objective optimization in image segmentation, offering detailed explanations and hands-on coding examples using real-world datasets. This comprehensive approach ensures that readers gain both theoretical knowledge and practical experience in tackling advanced classification challenges.

Chapter 5 explores deep learning approaches to multiclass and multi-label classification, providing a comprehensive overview of advanced techniques. We begin with multiclass classification using multilayer perceptron (MLP) and then delve into more sophisticated methods, including convolutional neural networks (CNNs) and recurrent neural networks (RNNs). The chapter also covers innovative approaches using deep generative networks, such as generative adversarial networks (GANs), and transfer learning. Readers will gain both theoretical insights and practical experience through hands-on coding examples, equipping them with the skills to implement deep learning models for complex classification tasks.

Chapter 6 summarizes the concepts presented throughout this book, focusing on applications and trends in multi-objective, multi-label, and multi-class classifications. It explores the social impacts of these advanced classification techniques across various domains, including health care, agriculture, entertainment, education, and finance. The chapter also addresses critical issues related to privacy and data security, highlighting the importance of ethical considerations and robust data protection measures in the deployment of these technologies. Through this comprehensive discussion, readers will gain an understanding of the broader implications and potential benefits of multi-objective and multi-label classification in real-world scenarios.

This book will explore the intricate world of data classification with "Multi-objective, Multi-class, and Multi-label Data Classification with tackling the problems related to class imbalance." This book explores sophisticated methods and strategies for working with complicated datasets, tackling the difficulties of various classes, many objectives, and complicated labeling tasks. This excellent resource fosters a deeper grasp of multidimensional data analysis in today's data-driven world by providing readers with the skills and insights needed to navigate the subtleties of modern classification jobs, from algorithmic techniques to practical applications. We believe this book will be benefited for the readers who are interested to work in this area in the future.

Dr. Sanjay Chakraborty
Postdoc Researcher
Department of Computer
and Information Science (IDA)
Linköping University
Linköping, Sweden

Associate Professor (lien)
Department of Computer Science
and Engineering
Techno International New Town
Kolkata, India

Dr. Lopamudra Dey
Associate Professor
Department of Computer Science
and Engineering
Meghnad Saha Institute of Technology
Kolkata, India

Postdoc Research Fellow
Department of Biomedical
and Clinical Sciences
Linköping University
Linköping, Sweden

Acknowledgements

We express our great pleasure, sincere thanks, and gratitude to the people who significantly helped, contributed, and supported the completion of this book. We are sincerely thankful to Principal of Techno International New Town, Kolkata, and Director of Meghnad Saha Institute of Technology Kolkata, India, for his encouragement, support, guidance, advice, and suggestions to complete this book. Our sincere thanks to Professor and Head, Department of Computer and Information Science, Linköping University, Sweden, and Dr. Anirban Mukhopadhyay, Professor, Department of Computer Science and Engineering, University of Kalyani, Kalyani, India, for their continuous support, advice, and cordial guidance from the beginning to the completion of this book.

We would also like to express our honest appreciation to our colleagues at the Linköping University, Sweden, Techno International New Town, India, and Meghnad Saha Institute of Technology Kolkata, for their guidance and support.

We are also very thankful to the reviewers for reviewing the book. This book would not have been possible without their continuous support and commitment toward completing the review on time.

To complete this book, the entire staff at Springer extended their kind cooperation, timely response, and expert comments and guidance, and we are very thankful to them.

Finally, we sincerely express our special and heartfelt respect, gratitude, and gratefulness to our family members and parents for their endless support and blessings.

To my parents, wife Lopamudra Dey, and son Arohan Dey Chakraborty for their love and inspiration.

—Dr. Sanjay Chakraborty

To my parents, husband Sanjay Chakraborty, and son Arohan Dey Chakraborty for their love and inspiration.

—Dr. Lopamudra Dey

Contents

About the Authors

Dr. Sanjay Chakraborty is currently associated with Department of Computer and Information Science (IDA), REAL, AIICS, Linköping University, Sweden, as Postdoc Researcher. He is also associated as an Associate Professor (lien) of the Department of Computer Science and Engineering, Techno International New Town, Kolkata, India. He did his B.Tech. from West Bengal University of Technology, India in the year 2009. He completed his Master of Technology (M.Tech.) from National Institute of Technology (NIT), Raipur, India, in the year of 2011. He completed his Ph.D. at AKCSIT, University of Calcutta in 2022. Dr. Chakraborty is a recipient of the University Silver Medal from NIT Raipur in 2011 for ranking first class second in M.Tech. He has 14 years of teaching and research experience. He has published over 70 research papers in various international journals, conferences, and book chapters. He is the author of three books published by Lap Lambert Germany, Springer DIR, and Springer EAI series, respectively. Dr. Chakraborty has five patents. His research interests include machine learning, applied AI, and quantum computing. He is Professional Member of IAENG and UACEE. Dr. Chakraborty is Active Member of the board of reviewers in various International Journals, Transactions, and Conferences. He is Recipient of "IEEE Young Professional Best Paper Award" in 2017. He has also achieved the top five best paper recognition by *Ain Shams Engineering Journal*, Elsevier, and most cited author award from *Biomedical Journal*, Elsevier in 2021. You can visit his homepage to know about his work and research at https://sites.google.com/view/portfolio-sanjay/home?authuser=4.

Dr. Lopamudra Dey is working as a Postdoc research fellow at Department of Biomedical and Clinical Sciences, Linköping University, Sweden and associated as an Associate Professor (lien) in the Department of Computer Science and Engineering in Meghnad Saha Institute of Technology, Kolkata, India. She completed B.Tech. from West Bengal University of Technology, Kolkata, India, in Computer Science and Engineering in 2009. She received a bronze medal in her bachelor's degree. In 2011, she completed M.Tech. from University of Kalyani, West Bengal, India. She obtained her Ph.D. in Computer Science from University of Kalyani, West Bengal, in 2021. Her areas of interests include computational biology, bioinformatics, data mining, and deep learning. She was awarded the "Most Cited Author" for the paper "Machine Learning Techniques for Sequence-based Prediction of Viral-Host Interactions between SARS-CoV-2 and Human Proteins" by the Biomedical journal, Elsevier in 2021. She had attended and presented at many national and international symposiums. She has published more than 35 research articles in journals, conferences, and books. She is also author of an international book published by Springer DIR series. You can visit her homepage to know about her work and research at https://scholar.google.co.in/citations?user=3DyQ8KYAAAAJ&hl=en.

Chapter 1
Introduction to Classification

In computer science, learning refers to the ability of a computer system to improve its performance on a task through experience. This kind of experience is frequently acquired through handling and evaluating data, detecting patterns, and formulating judgments or forecasts in response to these patterns. Essentially, computer science learning is creating models and algorithms that can automatically adjust or enhance their performance based on data, enabling them to carry out tasks without requiring explicit programming for every possible circumstance. Numerous applications, including as recommendation systems, driverless cars, natural language processing, image and audio recognition, and more, use learning algorithms. We will cover some fundamental information in this chapter that is helpful for studying supervised learning with an emphasis on multi-class, multi-label, and multi-objective learning.

1.1 Classification Overview

In the context of learning, classification is the process of grouping data according to certain characteristics into predefined classes or categories. To learn a mapping from input information to a discrete output label is a basic problem in supervised machine learning. A model is trained on labeled data—that is, data points with known class labels—in order to perform classification tasks. In order to forecast previously unseen or new occurrences and assign them to one of the specified classes, the model analyzes patterns and relationships found in the data. Support vector machines (SVM), random forests, logistic regression, decision trees, and neural networks are typical techniques used in classification [1]. The applications for classification are numerous and include image recognition, sentiment analysis, identification of spam, healthcare diagnosis, and more. The development of a model or classifier that can predict categorical labels for given input data is the aim of classification, which is in fact a crucial data analysis task. When classifying data, each data point is linked to a predetermined class label,

S. Chakraborty and L. Dey, *Multi-objective, Multi-class and Multi-label Data Classification with Class Imbalance*, Springer Tracts in Nature-Inspired Computing,
https://doi.org/10.1007/978-981-97-9622-9_1

and the input data is composed of a collection of features or attributes. The goal is to use a training dataset with known labels to learn a mapping from the input features to the associated class labels. As an example, a machine learning approach can be used to classify usual health data or facial images to determine whether a child has autism symptoms. This prediction can be done in two ways.

1. Using typical health features of a set of child samples.

Or

2. Using facial image features of a set of child samples.

First, a dataset containing health-related information, such as developmental milestones, medical history, and behavioral patterns, must be gathered. It may also include facial images. This dataset would include labeled examples that show whether or not each child has an autism spectrum disorder (ASD) diagnosis. To extract meaningful information from the data, feature selection and preprocessing would then be carried out; this may entail locating pertinent health indicators or removing face traits from images. After that, supervised learning methods like logistic regression, support vector machines, or deep neural networks can be used to train a classification model. The model would become adept at connecting data patterns to the existence or lack of autistic symptoms. For example, it could pick up on certain visual features or behavioral patterns linked to ASD. Performance criteria such as specificity, sensitivity, and accuracy would be utilized in the model evaluation process to gauge how well the classifier performs in predicting the existence of autistic symptoms based on the input data. The model can be used to determine whether a new child has autism symptoms based on their health information or face images once it has been trained and validated. With the use of this prediction approach, doctors may be able to diagnose and intervene in children with ASD earlier and provide better care and support. During training, the model learns patterns and relationships in the training data, which is composed of input features and corresponding labels or outcomes. Then, during testing or inference, the model applies what it has learned to new, unseen data to make predictions or classifications. The output of this process is the model's predictions or classifications for the test or production data. This feedback loop of training and testing is essential for the model to improve its accuracy and generalization capabilities. In supervised learning, the model learns from labeled data, where each input is associated with a corresponding output. The goal is to train a model that can predict the output for new, unseen inputs accurately. Classification tasks involve predicting discrete labels or categories, while regression tasks involve predicting continuous values. Examples like predicting species of iris flowers based on measurements demonstrate the application of supervised learning in real-world scenarios. Supervised learning techniques are widely used in various domains, including neural networks and decision trees, to create models that learn from labeled data and make predictions or classifications based on learned patterns and relationships [1]. The supervised method distinguishes between the observed data, or training data, which is often structured data that is provided to the model during training. The prediction model is constructed throughout this phase using

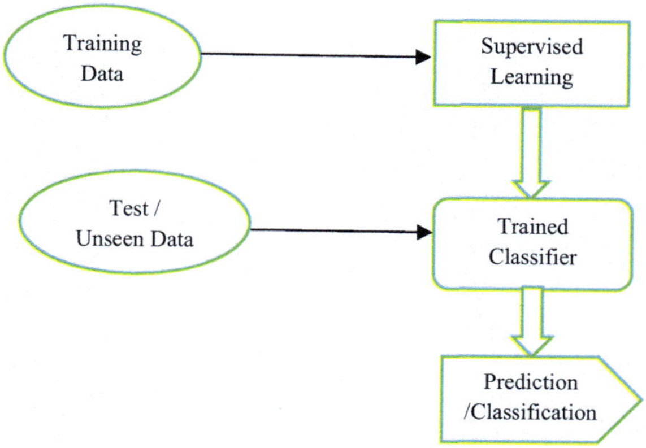

Fig. 1.1 Role of supervised learning

the supervised learning method. The fitted model would attempt to forecast the most likely labels for a fresh batch of samples (x) in the testing set after it had been trained. The role of Supervised learning is depicted in Fig. 1.1. Supervised learning can be categorized into the following categories according on the type of target,

1. The task of predicting y is referred to as classification if y has values within a predetermined set of category outcomes (integers).
2. Regression is the term used to describe the job of predicting y if it has floating point values.

In machine learning, the classification task is categorized based on the number of target classes in the output label. When there are only two distinct classes or outcomes to predict, such as yes/no, true/false, or autistic/non-autistic (according to the previous example), it is referred to as binary classification [2]. This scenario involves determining a decision between two mutually exclusive possibilities. In contrast, multi-class classification occurs when there are more than two possible classes or categories (high autistic, medium autistic, and low autistic) for the output label. Multi-class classification models are designed to assign probabilities across multiple classes and predict the most likely class for a given input. The choice between binary and multi-class classification significantly influences the selection of appropriate algorithms, evaluation metrics, and modeling strategies tailored to the specific nature of the problem being addressed [3, 4]. A brief identification of different types of classification is introduced in Table 1.1.

Table 1.1 Identification of classification types

b	No. of classes	No. of target labels	Objectives
Binary	2	1	1
Multi-class	> 2	1	1
Multi-label	> 2	None or All	1
Multi-objective	2 or > 2	1	> 1

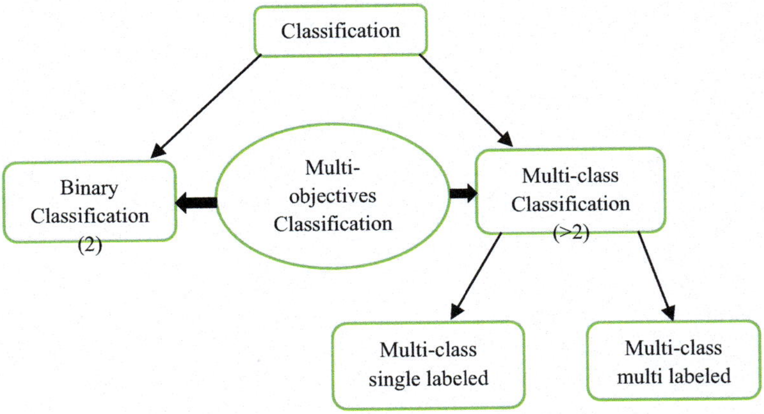

Fig. 1.2 Difference among binary, multi-class, and multi-labeled classifications

Figure 1.2 illustrates the distinctions between binary classification, multi-class classification, and multi-labeled classification within the context of machine learning tasks. In binary classification, the task involves predicting between two mutually exclusive classes or outcomes, such as "spam" or "non-spam," "positive" or "negative." Multi-class classification, on the other hand, deals with predicting among more than two distinct classes or categories for each instance, where each instance belongs to only one of these classes.

In contrast, multi-labeled classification allows instances to be associated with multiple labels or categories simultaneously, meaning each instance can have a set of characteristics or attributes assigned to it. Notably, multi-objective classification, which involves optimizing multiple objectives or criteria simultaneously, can be a component of either binary or multi-class classification tasks, depending on the specific context and requirements of the problem being addressed. This visual representation helps clarify the differences and relationships between these various types of classification tasks in machine learning. Classification is being employed globally in a variety of areas, including data mining, science, industry, and law. In terms of technology, machine learning has grown in importance and use. A few applications of machine learning include face recognition, self-driving cars, radiologists searching for early indications of cancer, and Chabot for online customer support. Other examples of its implementation are several security features like face recognition and access control. Classification may also be applied in the following cases:

1. Logical rules alone cannot adequately identify relevant rules.
2. Potential results are not identified in advance of the event.
3. More important than interpretability is precision.
4. The data is faulty and could pose issues for well-established analytical techniques.

1.2 Class Imbalance Problems

Although classification is useful in many fields, it also has many drawbacks, such as handling missing data, data identification that is irrelevant at the time of entry, data separation because of discrepancies with other known data, and equipment malfunctions and misunderstandings brought on by incomplete record entry. These issues can be resolved by replacing the entire set of omitted values with a known global constant. Certain institutions that employ machine learning encounter issues like data biases, untainted data, lack of control, and subpar output. Within loss of control environments, machine learning is unable to regulate the rate at which results are generated. Careful consideration and being prepared to apply the procedure are crucial for company. Filtering spam presents another challenge. We focus on providing a yes/no response when determining whether an email contains the most crucial information. In order to address these issues, we must develop a system that can automatically categorize fresh emails [5].

A significant volume of skewedly distributed data is generated in many real-time applications. If there is a larger sample from one class than the other, the dataset is considered highly skewed. When there is an imbalance in the dataset, the class with more instances is referred to as the major class, and the class with comparatively fewer occurrences is referred to as the minor class. Applications like the identification and prognosis of uncommon but serious diseases are far more significant than standard medical care. Situations are seen in different domains, such controlling risk, anticipating equipment breakdowns, identifying fraud in financial operations, and identifying network intrusions. Due to their bias toward the larger classes, the majority of classifiers in this scenario exhibit extremely low classification accuracy for the minor classes. Another possibility is that the classifier ignores the minor class and predicts everything as major class. Class imbalance is a common challenge in classification tasks where the distribution of class labels in the dataset is highly skewed, meaning that one class is significantly more prevalent than others. As a result, the classifier may become biased in favor of the majority class and perform poorly when it comes to predicting minority classes. A class imbalance may lead to misclassification and decreased performance since the model gives the dominant class more importance than the minority classes when it comes to accuracy. There are three main categories of approaches that have been proposed to address the issues related to class imbalance: algorithmic approaches, data preprocessing approaches, and feature selection approaches. Techniques like resampling (over or undersampling), using various evaluation metrics like precision, recall, and F1-score, or using sophisticated algorithms made to handle imbalanced datasets—like ensemble methods or cost-sensitive learning approaches—are frequently used to address class imbalance. In order to guarantee that classifiers produce precise and trustworthy predictions for every class, it is imperative to manage class imbalance. Figure 1.3 shows various types of class imbalance methods.

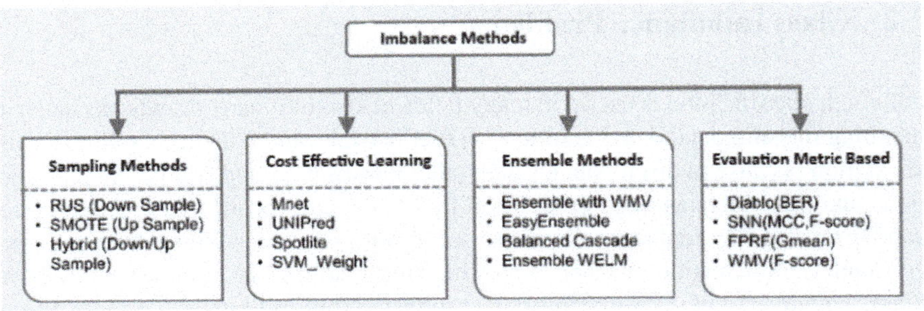

Fig. 1.3 Types of class imbalance methods

When classifying data, algorithms typically provide greater weight to accurately categorize the majority class samples. In some instances, misclassifying a rare occurrence might lead to more significant issues than a regular occurrence. When a cancerous cell is detected, for instance, misclassifying noncancerous cells may result in further clinical testing for medical diagnosis, but misclassifying harmful cells has extremely substantial health hazards. However, due to the design principles of most machine learning algorithms, which optimize overall classification accuracy and misclassify minority classes as a result, minority class examples are more likely to be misclassified in classification problems with imbalanced data than majority class examples [6].

1.3 Binary Classification

An essential job in statistics and machine learning is binary classification, which is predicting one of two possible outcomes for each instance given a set of data. Preprocessing the data involves encoding the categorical variables, handling missing values, and normalizing the features. Several methods, including logistic regression, neural networks, decision trees, support vector machines (SVM), random forests, and random forests, can be utilized to construct models that have the ability to recognize the fundamental patterns present in the data. Optimization methods such as gradient descent or tree-based algorithms make adjustments to the model while it is being trained [2]. A variety of measures are used to evaluate the model, such as accuracy, precision, recall, F1-score, and area under the receiver operating characteristic curve (AUC-ROC), which offer information about the model's performance in various contexts. Resolving class imbalance—a situation in which one class is noticeably more common than the other—is a frequent problem that can be solved with the help of ensemble methods, algorithmic modifications or resampling. A proper decision threshold must also be chosen because it dictates the trade-off between false positives and false negatives [7]. Binary classification is an essential tool in machine learning and data analysis, with applications spanning a wide range of fields such as image identification, sentiment analysis, medical diagnosis, fraud detection, spam detection, and customer churn prediction. Some key aspects of binary classification are described below.

(a) The goal of binary classification is to train a model that, given the attributes in the dataset, can correctly classify new examples.
(b) Binary classification datasets are made up of examples that are each characterized by a collection of characteristics (also known as independent variables or predictors) and a binary label (dependent variable) that corresponds to it. Features may be categorized, numerical, or a mix of the two.
(c) For binary classification, a variety of machine learning techniques can be applied, such as logistic regression, decision trees, random forests, support vector machines (SVM), and neural networks. From the labeled training data, the model learns patterns and correlations between the class labels and the features during training.
(d) A binary classification model's performance is usually assessed using a variety of measures after training, including area under the receiver operating characteristic curve (AUC-ROC), accuracy, precision, recall, F1-score, and area under the precision-recall curve (AUC-PR). These measures shed light on several elements of the model's performance, including its capacity to handle class imbalance, accurately categorize positive and negative examples, and produce accurate predictions [8].
(e) A decision threshold is employed in binary classification to translate the continuous output probabilities of the model into binary predictions. Depending on the particular requirements of the application, adjusting this threshold may involve a trade-off between precision and recall.

For binary classification tasks, a variety of model types can be applied; each has advantages and disadvantages as well as suitability for particular applications and data types. Below is a summary of a few typical types [2, 7–10]:

(a) **Logistic Regression**: This straightforward and popular approach is used for binary classification. A logistic function is used to transfer the input features to the likelihood of falling into one of the two classes, modeling the probability of the binary outcome. Logistic regression, although basic, can work well for datasets that can be divided linearly and can be used as a foundational model for more intricate algorithms.
(b) **Decision Trees**: To categorize instances, decision trees divide the feature space into regions and classify each node depending on its feature values. They are helpful for comprehending the model's decision-making process since they are simple to interpret and visualize. Nevertheless, decision trees are susceptible to overfitting, particularly when dealing with intricate datasets. To counteract this, one might employ strategies such as pruning or ensemble methods.
(c) **Support Vector Machines (SVMs)**: SVMs look for the hyperplane in the feature space that best divides instances of various classes from one another. They are efficient with datasets that have distinct margins of separation and

perform well in high-dimensional spaces. SVMs are flexible for handling a variety of data sources since they may be kernelized to handle nonlinear decision boundaries.

(d) **Random Forest**: Random forest is an ensemble learning technique that builds several decision trees during training and averages or votes among them to combine their predictions. It increases resilience and decreases overfitting, which addresses the drawbacks of individual decision trees. Complex datasets with nonlinear correlations between characteristics and labels are a good fit for random forest.

(e) **Gradient Boosting Machines (GBM)**: GBM produces weak learners (usually decision trees) one after the other in a sequential fashion and then combines them to produce a powerful prediction model. Iteratively adding new trees that fix mistakes in older ones, it minimizes a loss function. Because of their exceptional performance and adaptability, GBM algorithms like XGBoost, LightGBM, and CatBoost are very effective and frequently utilized in competitions and real-world applications.

(f) **Neural Networks**: Because deep learning models, in particular neural networks, can automatically recognize complicated patterns from unprocessed data, they have become increasingly popular for binary classification problems. Multiple layers of connected neurons make up neural networks, which may simulate extremely nonlinear interactions between features and labels. Depending on the type of data, neural network architectures such as feedforward neural networks, convolutional neural networks (CNNs), and recurrent neural networks (RNNs) can be modified for binary classification tasks.

(g) **Naïve Bayes**: When it comes to binary classification tasks, the Bayesian binary classifier employs the concepts of Bayesian inference. It uses the Bayes theorem to determine the likelihood that a particular instance, based on its attributes, belongs to each class. The Bayesian classifier is a systematic framework for classification that can handle uncertainty and incorporate previous information about the data by estimating class probabilities and making decisions based on these probabilities. It provides a probabilistic interpretation of classification judgements and is especially helpful when working with small datasets or when previous knowledge about the class distributions is available.

(h) **Nearest Neighbors**: The k-closest neighbors (kNN) algorithm, sometimes referred to as the nearest neighbors classifier, is a straightforward yet powerful nonparametric classification technique. The majority vote of the instances' closest neighbors in the feature space determines which classifies them. The technique finds the k-closest neighbors to a given instance by calculating the distances or similarities between them, usually using metrics like cosine similarity or Euclidean distance. It then predicts which majority class label among these neighbors will be present. Although kNN is simple to comprehend and computationally efficient, it can be costly to process huge datasets and has difficulty with high-dimensional feature spaces. Despite these drawbacks, kNN is still a flexible and natural method for classification tasks. A basic case study on binary classification is discussed below.

- **Case study on diabetes classification of patients:**
  - **Introduction:** Based on the patients' medical characteristics, we investigate in this case study the use of binary classification in predicting the onset of diabetes in patients. By prompt management and lifestyle changes, early identification of diabetes can greatly enhance patient outcomes.
  - **Problem Statement:** A chronic medical disease that is becoming more commonplace globally is diabetes. For preventive health care to be effective, early identification of people who are at high risk of developing diabetes is essential. The goal is to create a machine learning model that, using a patient's medical history and demographic data, can reliably forecast whether or not the patient will develop diabetes within a given time range.
  - **Data Collection:** The dataset consists of patient medical records with characteristics including age, gender, blood pressure, insulin, plasma glucose, body mass index (BMI), and diabetes status (target variable). In order to guarantee diversity and representativeness, data is gathered from several healthcare facilities and research projects.
  - **Data Preprocessing:** To manage missing values, normalize numerical features, and encode categorical variables, the dataset is put through a series of preprocessing stages. In order to guarantee that each feature contributes equally to the model's performance, feature scaling and normalization are used.
  - **Feature Selection and Engineering:** To determine the most pertinent features for forecasting the onset of diabetes, feature selection approaches such correlation analysis, feature importance ranking, and domain knowledge are used.
  - **Model Selection and Training:** Support vector machines (SVM), gradient boosting, random forest, logistic regression, and neural networks are among the binary classification algorithms taken into consideration. Using methods like cross-validation, these models are trained on the labeled dataset in order to optimize hyperparameters and guarantee reliable performance.
  - **Model Evaluation:** Standard evaluation metrics including accuracy, precision, recall, F1-score, and area under the ROC curve (AUC) are used to assess each model's performance. To guarantee accurate predictions, the model's calibration and discriminating abilities are also evaluated.
  - **Deployment and Integration:** The model is integrated into the healthcare system to provide real-time diabetes onset prediction after it has been trained and validated.
  - **Clinical Validation and Feedback:** To evaluate the deployed model's effectiveness in an actual healthcare environment, it is put through clinical validation. In order to pinpoint problem areas and enhance the model, input from patients and healthcare professionals is gathered.
  - **Outcomes and Clinical Implications:** The prediction model shows excellent precision and consistency in identifying those who are susceptible to diabetes. Healthcare practitioners can improve patient outcomes and lower the frequency of diabetes-related complications by prioritizing preventative

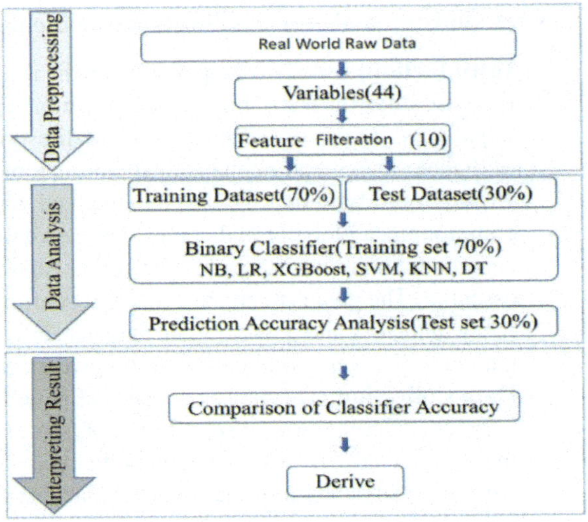

Fig. 1.4 Basic case study flow on binary classification

measures including lifestyle counselling, early screening, and customized treatment programmes by utilizing machine learning algorithms.

In the context of predicting the beginning of diabetes, this case study emphasizes the potential of binary classification in the prediction of medical diseases. Healthcare practitioners can use machine learning models to give proactive and individualized care by utilizing data-driven insights. This will ultimately improve patient outcomes and lessen the financial burden of treating chronic diseases like diabetes. Figure 1.4 illustrates that a basic case study on binary classification is divided into three stages, such as data collection and preprocessing, data analysis, and result interpreting [9, 10].

1.4 Multiclass Classification

In multi-class classification, the target variable consists of multiple distinct classes or categories, but each individual instance within the dataset is associated with only one of these classes. For example, in email classification, the target classes might be "spam" or "non-spam"; in loan default prediction, the classes could be "default" or "non-default"; and in medical diagnosis, the classes might represent conditions like "COVID-positive" or "COVID-negative." While the overall target variable encompasses multiple classes, each specific instance or record is uniquely assigned to one class based on its characteristics or features [4]. This contrasts with binary classification, where there are exactly two mutually exclusive classes for each instance (e.g., "spam" vs. "non-spam" or "default" vs. "non-default"). Multi-class classification tasks require models to predict the most likely single class out of the multiple

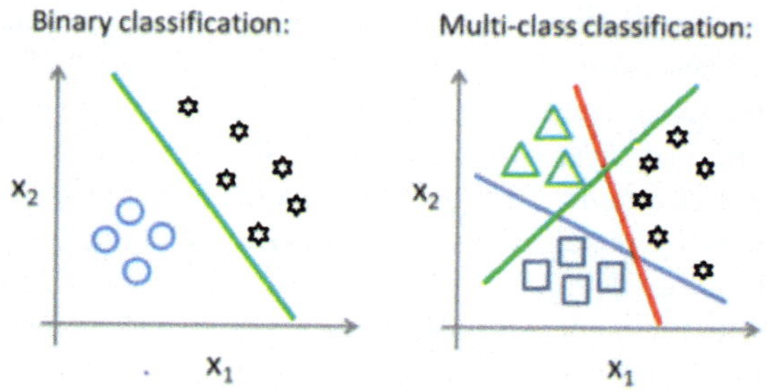

Fig. 1.5 Binary versus multi-class classification

possible categories for each instance, typically utilizing algorithms such as logistic regression, decision trees, random forests, or deep neural networks designed to handle multi-class scenarios effectively. Figure 1.5 represents the difference between binary and multi-class scenarios. Multi-class classification lacks the concept of normal and abnormal outcomes, in contrast to binary classification. Rather, samples are categorized as belonging to one of several recognized classes. On certain problems, there could be an extremely high number of class labels. In a facial recognition system, for instance, a model might identify a picture as being from one of thousands or tens of thousands of faces. Text translation models and other problems involving word sequence prediction could be categorized as a particular kind of multi-class categorization [11, 12].

Every word in the list of words that needs to be forecast has to be classified into many classes; the number of classes that can be predicted depends on the vocabulary size, which could include hundreds of words. For certain methods, the multiclass classification problem can be resolved by simply expanding the binary classification methodology. Neural networks, decision trees, Naive Bayes, k-nearest neighbor, and support vector machines are a few of them [13].

1.4.1 Neural Networks for Multi-class Classification

Multiclass problems are naturally extended by multilayer feedforward neural networks. We might have N number of binary neurons in the output layer, with binary output, as opposed to only one. It is the job of every output neuron to identify a specific class. For that class, this neuron should have an output code of 1, whereas the others should have a code of 0. As a result, $N = K$ neurons—where K is the number of classes—will be required in the output layer. The neuron that produces the highest output during testing an unknown sample is taken into account for determining the class label for that example.

A distinct binary codeword ranging from 0 to $2^N - 1$, where N is the total number of output neurons, is allocated to every class. The output codeword is compared to the codewords for the K classes when testing an unknown example, and the class that comes closest to the test-word based on some distance measure is declared the winner. In that situation, the number of distinct bits between the two codewords is known as the Hamming distance, and it is typically employed. Many deep learning models for classification are also available. A softmax function can be added to the final layer of almost any neural network to transform it into a classifier. The output vector of length K is generated by the softmax function, which also generates a probability distribution across K classes. The likelihood that the input falls into the appropriate class is represented by each element of the vector. The vector with the highest probability is selected, and its index is used to determine the most likely class. Although there are numerous neural network topologies available, some perform better than others. Convolutional neural nets (CNNs) do exceptionally well on multiclass classification tasks, particularly when dealing with text and images. Utilizing features that are invariant to scaling, transformation, and rotation, a CNN is able to extract valuable information from data. This improves its ability to identify potentially rotated, shrunken, or off-center images, which boosts its accuracy in image multiclass classification tasks [14].

1.4.2 Support Vector Machine (SVM) for Multi-class Classification

Although enhancements have been developed to handle the multiclass classification issue as well, the fundamental support vector machine (SVM) only allows binary classification. These additions address the separation of the several classes by adding more constraints and parameters to the optimization issue. Using a single (or collection of) hyperplane, an SVM divides the input into distinct classes by projecting it into a higher dimensional space. One support vector machine (SVM) is capable of binary classification and class differentiation. To distinguish between N classes, $(N - 1)$ SVMs can be used. Everyone would guess that they would be in one of the N classrooms.

1.4.3 Decision Tree for Multi-class Classification

Nodes in a decision tree for multi-class classification demonstrate features, branches decision-making, and leaves class labels. It is a hierarchical structure. In order to minimize impurity, the algorithm chooses the feature at each node that best divides the classes. Recursively, this process runs until a halting condition is satisfied. After that, the tree can be used to classify data that has not been seen by going from the root to a leaf and assigning the leaf's most common class. It is a flexible and understandable approach to complicated multi-class categorization problems. Either of the K classes in question may be referred to by the leaf nodes [13].

1.4.4 Naïve Bayes for Multi-class Classification

Based on Bayes' theorem, Naïve Bayes is a probabilistic algorithm used for multi-class categorization. It is very successful, especially for text categorization jobs, despite its simplicity. "Naïve" refers to the assumption of independence between features given the class designation. For a given instance, the algorithm determines the likelihood of each class and chooses the class with the highest probability to be the predicted class. Even with little datasets, it performs well computationally. Its performance in situations where features are highly correlated may be impacted by the possibility that its feature independence assumption is not always met. All things considered, Naïve Bayes is a strong and popular method for multi-class classification tasks, especially when there are a lot of features and a large training dataset.

Using binary classifiers, it is possible to decompose the multiclass classification problem into multiple binary classification tasks that can be completed quickly. The support vector machines are the most reliable and popular binary classifiers. The concept is comparable to assigning codewords to every class and then employing a variety of binary classifiers to solve a series of binary classification tasks, the solutions to which can establish the class label for novel data. Numerous techniques have been offered for this kind of decomposition [11, 12, 15].

(a) **One-versus-All**: The most straightforward method is to break down the classification problem into K binary problems, each of which separates a particular class from the remaining $K - 1$ classes. $N = K$ binary classifiers are needed for this method, in which the kth classifier is trained using positive examples from class k and negative examples from the remaining $K - 1$ classes. The classifier that generates the highest output during testing an unknown case is deemed to be the winner, and this class label is applied to that sample.

(b) **All-versus-All**: This method compares every class to every other class. Each pair of classes is discriminated against by a binary classifier, which eliminates the remaining classes. It builds $\frac{K(K-1)}{2}$ binary classifiers. A vote is conducted among the classifiers during the testing of a new example, and the class with the most votes wins. Research indicates that this strategy works better than the one-versus-all strategy overall.

- **Case study on handwritten characters recognition:**

We investigate the use of multiclass classification in handwritten digit recognition from photographs in this case study (Fig. 1.6). Optical character recognition (OCR) and digital document processing are two uses for handwritten digit recognition, which is a basic pattern recognition problem [13].

- **Problem Statement**: The goal is to create a machine learning model that, using the pixel values of grayscale photographs, can accurately detect handwritten numerals (0–9). The model should be highly accurate in classifying all digit classes and have good generalization to new data.

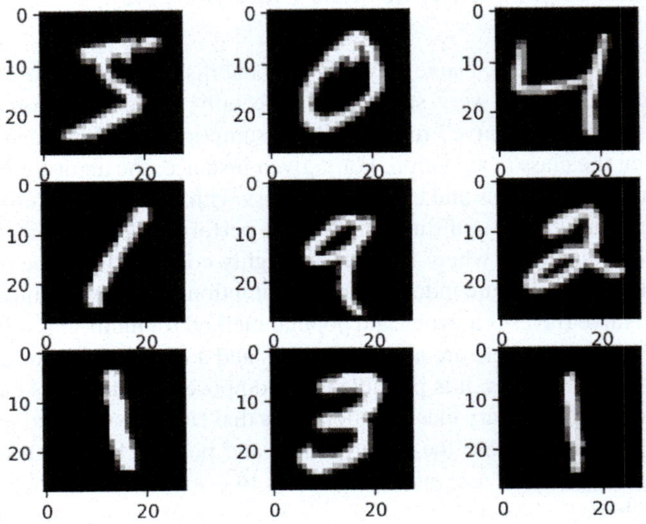

Fig. 1.6 Plot of the nine images in the MNIST dataset

- **Dataset Description**: The MNIST dataset, a popular benchmark dataset for handwritten digit recognition, was employed for this investigation. It has 10,000 test photos and 60,000 training images, each measuring 28 by 28 pixels. Because each image is labeled with the digit it represents (0–9), it can be used for challenges involving multiclass categorization.
- **Preprocessing of the Dataset**: The dataset is first normalized to scale pixel values between 0 and 1, images are reshaped into a flat vector format ($28 \times 28 = 784$ dimensions), and class labels are one-hot encoded to represent them as binary vectors before the model is trained.
- **Model Selection and Training**: A number of machine learning techniques, such as support vector machines (SVM), decision trees, random forests, logistic regression, and neural networks, are taken into consideration for multiclass classification. In order to maximize performance, these models are trained on the preprocessed training data using methods like cross-validation and hyperparameter tweaking.
- **Model Evaluation**: Metrics including accuracy, precision, recall, F1-score, and confusion matrix are used to assess each model's performance. Furthermore, methods such as grid search and k-fold cross-validation are utilized to guarantee the models' resilience and capacity for generalization.
- **Deployment and Integration**: For real-time digit recognition, the trained and validated model is integrated into production systems. Users can enter handwritten numbers and instantly receive precise predictions by integrating with programmes like mobile apps or OCR software.

- **Outcomes and Uses**: Across all classes, the implemented model recognizes handwritten numerals with excellent accuracy. It can be used to increase productivity and accuracy in a variety of applications, such as automated form recognition, bank cheque processing, automated postal automation, and digitalized document processing.

This case study illustrates how well machine learning techniques work with multi-class classification to identify handwritten numbers. For a variety of real-world uses, precise and effective digit recognition models can be created by utilizing the MNIST dataset and suitable preprocessing methods.

1.5 Multi-label Classification

Multi-label classification is indeed a distinct type of machine learning task where each instance or record in the dataset can be associated with multiple labels or categories simultaneously. This contrasts with traditional single-label classification, where each instance is assigned to one specific class from a set of mutually exclusive categories. In multi-label classification scenarios, a single instance can exhibit characteristics that correspond to multiple labels. For example, in movie genre classification, a film can be labeled as both "comedy" and "romantic" simultaneously [16]. Similarly, in personality prediction from text, an individual's traits might include being "introverted," "serious," "extroverted," "outgoing," "stress-free," "honest," and "sincere," with each characteristic represented as a separate label. Multi-label classification algorithms are designed to predict a set of labels or categories for each instance, making them suitable for applications where instances can possess multiple distinct attributes or characteristics [17]. Figure 1.7 represents the difference between multi-class and multi-label classification scenarios. These algorithms often involve adapting traditional classification methods and leveraging techniques that can handle the complexity of multi-label assignments efficiently. Multi-label classification problems are frequently modeled using a model that predicts several outputs, each of which is assumed to have a Bernoulli probability distribution. In essence, this is a model that predicts several binary classifications for every case. It is not possible to directly use classification methods designed for binary or multi-class classification to multi-label classification [18]. It is possible to employ specialized versions of common classification algorithms, also known as "multi-label versions," such as

- Decision trees with multiple labels
- Multi-label random forests
- Multi-label gradient boosting
- Classifier chains
- Label Powerset
- Multi-label k-nearest neighbors (MLkNN).

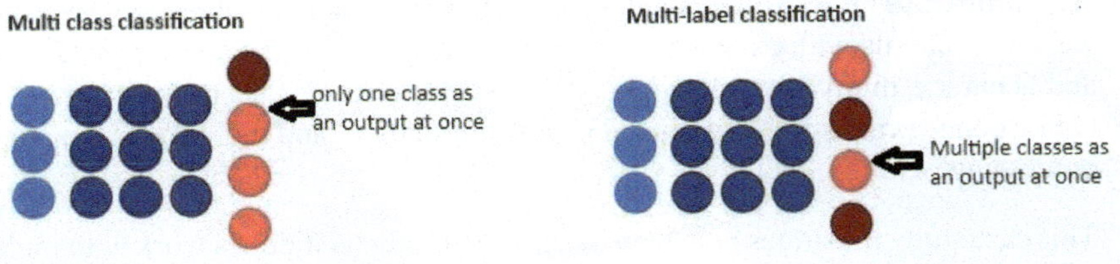

Fig. 1.7 Multi-class versus multi-label classification

Multi-label classification is not only used for data classification but also plays an interesting role for image classification. Multi-label active learning algorithms have become essential tools in the field of image classification, helping to improve classification accuracy and streamline the annotation process. These methods combine multi-label classification frameworks with active learning procedures to select the most informative subsets of unlabeled images for annotation iteratively. These algorithms reduce annotation costs and speed up the learning process by dynamically selecting photos that are best for model refinement through the use of uncertainty sampling, query-by-committee, or other active learning techniques. Moreover, they take into account the existence of many labels in each image, so accommodating the intrinsic complexity of multi-label classification problems. These techniques efficiently tackle the difficulties presented by large-scale image datasets with numerous labels per image through iterative improvement of the classification model based on the labeled data obtained during active learning phases. Multi-label active learning algorithms present a viable method for enhancing the effectiveness and precision of image classification systems across diverse fields, such as multimedia content analysis, medical imaging, and remote sensing [16]. A specific instance of data classification, where classes are hierarchically built and objects can be allocated to numerous paths of the class hierarchy at the same time, is one of the most difficult machine learning issues. Hierarchical multi-label classification (HMC) is the name given to this activity, which finds applications in bioinformatics problems like protein function estimation, image annotation, and text categorization. A new neural network architecture for hierarchical machine learning (HMC) called HMCN is proposed in a research paper. It can optimize both local and global loss functions at the same time to find global information from the entire class hierarchy and local hierarchical class relationships while penalizing hierarchical violations [19].

- **Case study on document tagging multi-label classification:**

The amount of documents being generated in today's digital age is massive, spanning multiple domains. Sorting, organizing, and retrieving these papers effectively all depend on how well they are classified. This job can be effectively solved with multi-label classification approaches, which enable documents to be assigned numerous tags at the same time. In this case study, we investigate the use of multi-label classification algorithms for content-based automatic document tagging [20].

- The **problem statement** states that our objective is to create a machine learning model that can tag documents from a predetermined set of categories appropriately. This is a multi-label classification problem since each document might be in one or more categories at the same time. Our goal is to anticipate suitable tags for unseen documents with a high degree of accuracy.
- **Data gathering and preprocessing**: We gather a sizable dataset of documents and their corresponding tags from a variety of sources. Preprocessing is done on the papers to get rid of extraneous letters, punctuation, and stopwords. Text normalization methods reduce inflectional forms to a single base form, such as stemming and lemmatization. Next, the dataset is divided into test, validation, and training sets.
- **Feature engineering**: To represent the content of each document, we extract pertinent features from it. Word embeddings (such as Word2Vec and GloVe), character-level embeddings, and term frequency-inverse document frequency (TF-IDF) vectorization are examples of common approaches. We might also include metadata elements like author information, publication date, and document length.
- We train the above suitable models using the training data and optimize their hyperparameters using cross-validation on the validation set. Evaluation metrics such as Hamming loss, F1-score, and exact match ratio are used to assess the performance of each model.
- We assess the models' performance on the test set after training in order to gauge how well they can generalize. When necessary, we adjust hyperparameters or feature representations to further enhance performance based on our analysis of the model's predictions.

1.6 Multi-objective Classification

Machine learning typically has to accomplish a number of goals, many of which are incompatible with one another. For instance, in feature selection, there are competing goals when it comes to maximizing feature quality and limiting the quantity of features. It is also well acknowledged that model complexity and approximation or classification accuracy must be traded off in the process of choosing a model. Conventional learning techniques reduce multi-objective machine learning issues to single-objective problems by aggregating many objectives into a scalar cost function [21]. The remarkable advancements in evolutionary multi-objective optimization have sparked a growing interest in the application of Pareto-based multi-objective optimization to machine learning. It has been demonstrated that the multi-objective machine learning methodology is especially effective at: (1) raising the standard single-objective machine learning methods' performance, (2) Producing extremely varied numerous Pareto-optimal models for building ensembles, and (3) attaining a desired balance between the interpretability and accuracy of fuzzy systems or neural networks [22, 23]. The following are the primary areas of focus for multi-objective machine learning:

- Feature extraction, feature selection, and multi-objective clustering.
- Selection of multi-objective models to enhance learning models' performance, including fuzzy systems, decision trees, neural networks, and support vector machines.
- Multi-objective model selection to enhance fuzzy systems' interpretability or learning models' interpretability, such as when removing symbolic rules from neural networks.
- Multi-criteria ensemble generation.

Multi-objective classification refers to the task of simultaneously optimizing multiple objectives or criteria in the process of classifying instances into multiple classes or categories. Multi-objective classification takes into account several competing objectives that may need to be prioritized or balanced, in contrast to traditional classification issues where a single target, such as accuracy, is optimized. Metrics such as accuracy, precision, recall, F1-score, and area under the ROC curve (AUC) could be included in these goals. The aim is to identify a set of alternatives, referred to as Pareto-optimal solutions that show a trade-off between these goals and where achieving one goal might result in a deterioration of another. Multi-objective classification techniques sometimes entail ensemble approaches or the optimization of a variety of classifiers to strike a balance between many performance criteria. When making decisions using several criteria at once, as in healthcare diagnosis, financial risk assessment, or cybersecurity anomaly detection, multi-objective categorization is especially helpful. Several target objectives, $L_1,\ldots, L_N$, describe various hypothesized qualities of interest in multi-objective learning. When these are estimated from a (single) dataset, $L'_1, \ldots, L'_N$ are the empirical objectives that result. When it comes to machine learning models, other pertinent quantities like efficiency, robustness, or fairness can also be expressed using a principled framework in the multi-objective setting, unlike in the single-objective scenario where the objective function is almost always linked to a prediction quality measure. As a result, we permit the objectives to take on alternative shapes in addition to expected values over per-sample loss functions, and we do not limit their empirical estimations to averages across samples. The hypotheses are not completely organized by the multi-objective setup. As a result, there will no longer be an overall best hypothesis a priori [24].

- If a hypothesis, $h \in H$, is Pareto-optimal when it comes to the multi-objective optimization problem of reducing $L'_1, \ldots, L'_N$ (where the coefficients are calculated from a training set), then it is considered empirically Pareto-optimal.

The machine learning community has been paying more and more attention to multi-objective learning in the past few years, and numerous techniques have been suggested for it. Multi-objective generalization is impossible until at least single-objective generalization maintains, as the multi-objective scenario rigorously generalizes the single-objective one [25].

- **Case study on multi-objectives classification in health care**:

Accurately identifying complicated diseases, like cancer, which have several subtypes and diverse patient groups, is frequently difficult for healthcare professionals. Conventional binary classification models could find it difficult to prioritize competing diagnostic goals like sensitivity and specificity and to adequately represent the variety of features present in these disorders. Multi-objective classification may accurately identify cancer subtypes while balancing sensitivity and specificity, which is crucial for customized therapy planning. It also helps in the discovery of biomarkers which is important to many aims, including early detection, prognosis, and therapy response.

- **Goal**: To improve clinical decision-making and diagnostic accuracy by creating a multi-objective classification model for disease diagnosis that simultaneously improves sensitivity, specificity, and accuracy.
- **Data Gathering and Preprocessing**: Patients with the target disease as well as controls provided clinical and genetic data for the dataset. Gene expression profiles, medical histories, test findings from labs, and demographic data are among the features. Preprocessing includes feature selection to lower dimensionality, encoding categorical variables, addressing missing values, and normalizing numerical features.
- **Model Construction**: Establish the goals of the multi-objective classification model, such as maximizing overall accuracy while minimizing false positives and false negatives by optimizing specificity and sensitivity.
- **Selecting an Algorithm**: Pick multi-objective evolutionary algorithms (MOEAs), multi-objective genetic programming (MOGP), or ensemble techniques based on Pareto optimization are examples of algorithms that can handle multi-objective optimization.
- **Model Training**: Using MOEAs or ensemble techniques, train several classifiers and simultaneously optimize them for a variety of performance criteria. To assess model performance and determine Pareto-optimal solutions that represent the trade-off between several objectives, apply cross-validation.
- **Model Evaluation:** Utilize performance metrics like F1-score, area under the ROC curve (AUC), recall, sensitivity, specificity, accuracy, and precision to assess the trained classifiers.

Through the simultaneous optimization of many diagnostic objectives, multi-objective classification in machine learning provides an effective tool for disease diagnosis in the healthcare industry. Multi-objective classifiers can boost diagnosis accuracy, enable personalized therapy, and support clinical decision-making in complex disease situations by incorporating a variety of performance metrics into the model development process.

1.7 Conclusions

In this chapter, we have provided a comprehensive introduction to supervised classification, covering various learning approaches. The overview of classification techniques has provided a comprehensive understanding of the fundamentals and methodologies employed in supervised learning. We explored binary, multi-class, multi-label, and multi-objective classification methods, offering insights into their respective characteristics and applications. Additionally, we have presented several engaging case studies to illustrate real-world implementations of these classification techniques. Through our discussion, readers have gained a holistic understanding of supervised classification, its diverse methodologies, and its practical significance in solving complex classification problems across different domains.

Bibliography

1. Loog M (2018) Supervised classification: quite a brief overview. In: Machine Learning Techniques for Space Weather, pp 113–145
2. Kumari R, Srivastava SK (2017) Machine learning: a review on binary classification. Int J Comput Appl 160(7)
3. Khan MS, Nath TD, Hossain MM, Mukherjee A, Hasnath HB, Meem TM, Khan U (2023) Comparison of multiclass classification techniques using dry bean dataset. Int J Cogn Comput Eng 4:6–20
4. Panca V, Rustam Z (2017) Application of machine learning on brain cancer multiclass classification. In: AIP Conference proceedings, vol 1862, No. 1. AIP Publishing
5. Ganganwar V (2012) An overview of classification algorithms for imbalanced datasets. Int J Emerg Technol Adv Eng 2(4):42–47
6. Cateni S, Colla V, Vannucci M (2014) A method for resampling imbalanced datasets in binary classification tasks for real-world problems. Neurocomputing 135:32–41
7. Bellinger C, Sharma S, Japkowicz N (2012) One-class versus binary classification: which and when? In: 2012 11th International conference on machine learning and applications, vol 2. IEEE, pp 102–106
8. Galar M, Fernández A, Barrenechea E, Bustince H, Herrera F (2011) An overview of ensemble methods for binary classifiers in multi-class problems: Experimental study on one-vs-one and one-vs-all schemes. Pattern Recogn 44(8):1761–1776
9. Kanvinde N, Gupta A, Joshi R, Gerela P (2023) Binary classification for high dimensional data using supervised non-parametric ensemble method. In: Intelligent Computing and networking: proceedings of ICICN 2022. Springer Nature, Singapore, pp 15–23
10. Nellore SB (2015) Various performance measures in Binary classification—an overview of ROC study. IJISET-Int J Innov Sci Eng Technol 2(9):596–605
11. Alsafy BM, Aydam ZM, Mutlag WK (2019) Multiclass classification methods: a review. Int J Adv Eng Technol Innov Sci 5(3)
12. Fu L, Liang P, Li X, Yang C (2021) A machine learning based ensemble method for automatic multiclass classification of decisions. In: Proceedings of the 25th International conference on evaluation and assessment in software engineering, pp 40–49
13. Chaudhary A, Kolhe S, Kamal R (2016) An improved random forest classifier for multi-class classification. Inform Process Agric 3(4):215–222
14. Bujang SDA, Selamat A, Ibrahim R, Krejcar O, Herrera-Viedma E, Fujita H, Ghani NAM (2021) Multiclass prediction model for student grade prediction using machine learning. IEEE Access 9:95608–95621

15. Vidyarthi A, Agarwal R, Gupta D, Sharma R, Draheim D, Tiwari P (2022) Machine learning assisted methodology for multiclass classification of malignant brain tumors. IEEE Access 10:50624–50640
16. Wu J, Sheng VS, Zhang J, Li H, Dadakova T, Swisher CL et al (2020). Multi-label active learning algorithms for image classification: overview and future promise. ACM Comput Surv (CSUR) 53(2):1–35
17. Tsoumakas G, Katakis I (2007) Multi-label classification: an overview. Int J Data Warehousing Mining (IJDWM) 3(3):1–13
18. Herrera F, Charte F, Rivera AJ, Del Jesus MJ, Herrera F, Charte F et al (2016) Multilabel classification. Springer International Publishing, Berlin, pp 17–31
19. Wehrmann J, Cerri R, Barros R (2018) Hierarchical multi-label classification networks. In: International conference on machine learning. PMLR, pp 5075–5084
20. Tarekegn AN, Giacobini M, Michalak K (2021) A review of methods for imbalanced multi-label classification. Pattern Recogn 118:107965
21. Gong Z, Chen H, Yuan B, Yao X (2018) Multiobjective learning in the model space for time series classification. IEEE Trans Cybern 49(3):918–932
22. Xue Y, Tang Y, Xu X, Liang J, Neri F (2021) Multi-objective feature selection with missing data in classification. IEEE Trans Emerg Topics Comput Intell 6(2):355–364
23. Cai W, Chen S, Zhang D (2009) A multiobjective simultaneous learning framework for clustering and classification. IEEE Trans Neural Networks 21(2):185–200
24. Jiao R, Nguyen BH, Xue B, Zhang M (2023) A survey on evolutionary multiobjective feature selection in classification: approaches, applications, and challenges. IEEE Trans Evolution Comput
25. Bai L, Li H, Gao W, Xie J, Wang H (2023) A joint multiobjective optimization of feature selection and classifier design for high-dimensional data classification. Inf Sci 626:457–473

Chapter 2
Class Imbalance and Data Irregularities in Classification

2.1 Introduction

The modern world has become data-driven, with the explosion of AI and machine learning leveraging this data to forecast corporate results, comprehend society, and promote growth and decision-making. Classification is among the most popular and extensively utilized supervised machine learning techniques. It serves as an aid in classifying data and has a wide range of uses, including speech recognition, image classification, fraud detection, medical diagnostic testing, email spam detection, etc. Before being applied to new, unknown data for prediction, the model in classification is thoroughly trained using the training set. It is then assessed on test data. For example, an algorithm can be trained to determine if an email is spam or legitimate. However, when anomalies of various kinds and imbalance classes are present in the training dataset, the performance of most well-known classifiers can significantly deteriorate. Conventionally, the machine learning algorithms assume certain criteria for training datasets. For example, equal representation of every class, for each characteristic in the dataset, there is a known value for each data instance, and across all the classes, the class-conditional distributions are similar. Data irregularities and class imbalance problems arise when violations of these ideal conditions impede a classifier's normal learning process. The majority of the real-world dataset suffers from these problems and generates erroneous judgments, flawed analysis, and insufficient operational performance. Therefore, to limit the impact of class imbalance and data abnormalities, addressing these difficulties needs careful consideration of preprocessing strategies, feature engineering, and model selection. In this chapter, we will discuss different aspects of class imbalance and data abnormalities problems and also the techniques for assessing model performance accurately in these situations. This chapter also discusses different solutions of handling them.

S. Chakraborty and L. Dey, *Multi-objective, Multi-class and Multi-label Data Classification with Class Imbalance*, Springer Tracts in Nature-Inspired Computing,
https://doi.org/10.1007/978-981-97-9622-9_2

2.2 Class Imbalance

When instances in a dataset are not evenly distributed among distinct classes, it is referred to as class imbalance. Stated differently, there are noticeably fewer examples in one or more classes than in other classes. In many machine learning and data analysis contexts, class imbalance is a prevalent problem that can affect the dependability and performance of models [1]. For example, the training dataset below (Fig. 2.1) presents a binary classification problem,

Total number of observations: 100
There are two target variables "Yes" or "No".

Instance-1

A dataset is considered imbalanced if there are 90 "Yes" responses and 10 "No" responses because the distribution of the two classifications is very uneven.

Instance-2

A balanced dataset is shown by 55 "Yes" responses and 45 "No" responses, as this indicates a roughly equal proportion of the two classes.

2.2.1 *The Problem with Class Imbalance in Machine Learning*

One class may dominate the others in many real-world datasets, creating a skewed distribution [2]. Several difficulties and issues with machine learning models may arise from this imbalance,

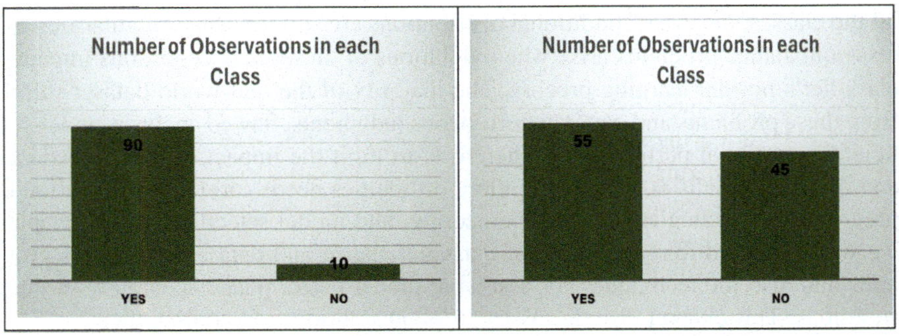

(a) (b)

Fig. 2.1 **a** Classification of the imbalanced dataset, **b** classification of the balanced dataset

(a) Favoritism for the Majority Class: In circumstances with a class imbalance in the dataset, the machine learning algorithms still achieve a high degree of accuracy by simply forecasting the majority class. However, this approach will not account for the minority class, which is typically the main reason for developing the model in the first place. A decision tree, logistic regression, or support vector machine (SVM), for instance, is a basic classification model that will not be able to detect the minor class data points if the class distribution indicates that 99% of the data belongs to the majority class.
(b) Inadequate Synopsis: When it comes to minority classes, in particular, imbalanced datasets might result in poor generalization to new, unknown data. The model may lack resilience since there are not enough examples for it to understand the main patterns of the minority class.
(c) Interpreting Feature Importance with Difficulty: Even if features that discriminate between the majority and minority classes are not intrinsically meaningful, models may give them more weight in imbalanced datasets. This can make determining a feature's actual significance difficult.
(d) Classification metrics: Selecting the appropriate assessment metrics is essential when working with imbalanced datasets in machine learning so that the model's performance may be reliably determined. For instance, the model's initial accuracy might be 99% in a dataset that contains 1000 photographs of sparrows and 99,000 images of crows. This statistic, however, might not adequately reflect the model's capacity to categorize the minority class (sparrows).

2.2.2 The Class Imbalance Problem in Deep Learning

In deep learning, the term "class imbalance problem" describes a situation in which there is a considerable skew in the training data's class distribution, favoring one or more classes over others. Due to their low representation in the dataset, the model may find it difficult to learn from the minority class or classes, which can cause a number of difficulties and issues throughout the training process [3]. The following are some of the main problems related to the deep learning class imbalance issue,

A. Deep learning models may prioritize learning patterns from the majority classes while ignoring the minority classes as a result of their tendency to optimize toward minimizing total error. Consequently, the model might not accurately categorize instances that belong to the minority classes.
B. When applied to unseen data, models trained on unbalanced datasets may perform poorly in terms of generalization, especially on minority class cases. This could be as a result of their inadequate schooling, which left them lacking in knowledge of the minority class or classes.
C. With imbalanced datasets, standard evaluation criteria like accuracy might be deceptive. For example, even if a classifier correctly predicts the majority class in every instance, it may nevertheless receive a high accuracy score if it misclassified instances of the minority class.

D. Training problems with loss function optimization might arise from imbalanced datasets. The model's actual performance might not be fully reflected in the gradients generated from unbalanced data, which could result in less-than-ideal convergence in training.

2.2.3 *Applications of Class Imbalance Problems*

Class imbalance issues arise in a wide range of fields and applications. The following instances highlight the difficulties posed by imbalanced datasets:

(a) Customer Churn Prediction: Retaining customers is one of the most crucial problems for businesses. Preventing customer turnover is a top priority when using a customer relationship management (CRM) strategy [2]. Large corporations use churn prediction models to identify potential churners before they quit. Accurately detecting possible churn cases is challenging due to the unbalanced nature of the data.
(b) Biomedical Research: The majority of real-world biomedical datasets often include high-dimensional features and small sample sizes. This has a negative impact on the model's classification performance and results in inaccurate disease diagnosis guidance. It is difficult to investigate an efficient classification technique for limited and unbalanced biomedical datasets [3].
(c) Fraud Identification: In financial statistics, fraudulent transactions are typically uncommon in comparison with legitimate ones. Accurately detecting fraud is essential, and the issue of class imbalance may have an impact on how well fraud detection algorithms work.
(d) Finding anomalies:
    Finding anomalies in a variety of domains, such as industrial processes, system logs, or network traffic, frequently requires working with unbalanced datasets in which the number of normal occurrences greatly exceeds the number of abnormal ones.
(e) Protection and Identifying Intrusions: Another domain where unbalanced datasets are prevalent is the identification of security concerns or network breaches. Since most network activity is benign, it can be difficult to identify the comparatively uncommon occurrences of malevolent behavior.
(f) Identifying Objects in Pictures:

Class imbalance in object detection tasks can result from certain classes being less common than others in image collections.

2.2.4 *Methods to Address Class Imbalance Problem*

Building efficient machine learning models requires addressing the issue of class imbalance, particularly in situations when one class is noticeably underrepresented. A few popular approaches to solving the issue of class imbalance are given below:

Resampling: Resampling is a commonly used class imbalance approach for handling very imbalanced datasets. Based on the real data, resampling creates a distinct sampling distribution (Fig. 2.2). Monte Carlo estimation or bootstrapping are other names for resampling. There is no set sample size needed for resampling. Consequently, the confidence intervals produced by the resampling procedure are more reliable the larger the sample [4].

Oversampling: Using this method, we replicate to raise the minority class sample size and attempt to match it with the majority class sample size [5]. This will improve the model's ability to forecast more accurately and without favoring the majority class. Oversampling can be described with the help of following example (Fig. 2.3),

Example

Let us use a similar thousand-record skewed training dataset.

Oversampling (Before):

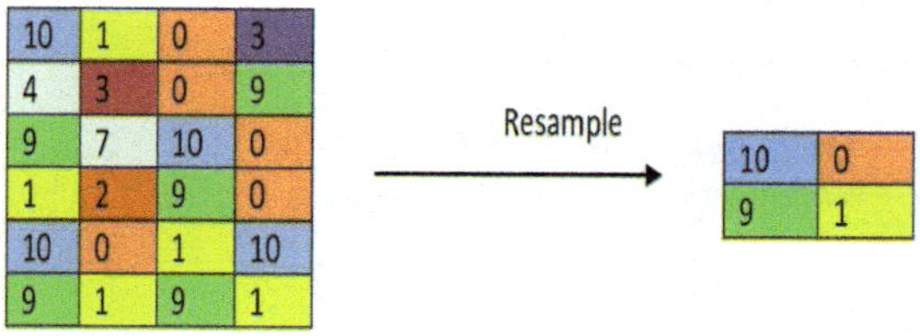

Fig. 2.2 Resampling method

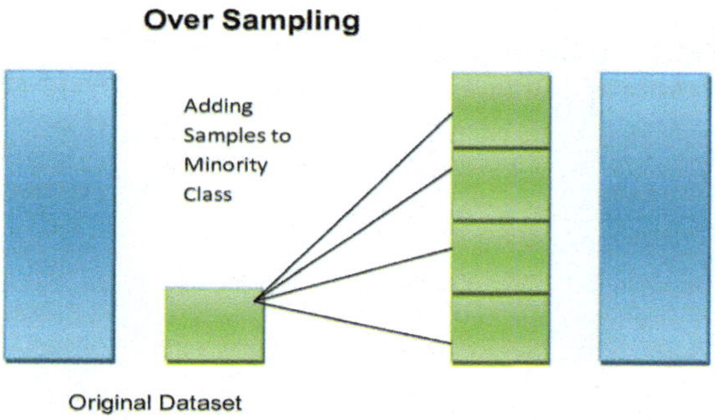

Fig. 2.3 Oversampling scenario

"Yes" target class = 800 instances
"No" target class= 200 instances

Oversampling (After):

"Yes" target class = 800 instances
"No" target class= 800 instances

Advantages:

The performance of the model is improved when patterns are preserved.

Disadvantages:

- Overfitting may result from data replication.
- It requires a lot of processing power.

Undersampling: Undersampling is a machine learning strategy that matches the size of the minority class with the majority class in order to balance the dataset's class distribution. This will improve the model's ability to forecast more accurately and without favoring the majority class [6]. A subset of the majority class instances equal to the size of the minority class would be chosen at random or using a different sampling strategy in order to undersample the dataset. Undersampling can be described with the help of following example (Fig. 2.4),

Example

Suppose, we have an imbalanced dataset with 1000 instances for training.

Undersampling (Before):

"Yes" target class = 800 instances
"No" target class= 200 instances

Undersampling (After):

"Yes" target class = 200 instances
"No" target class= 200 instances

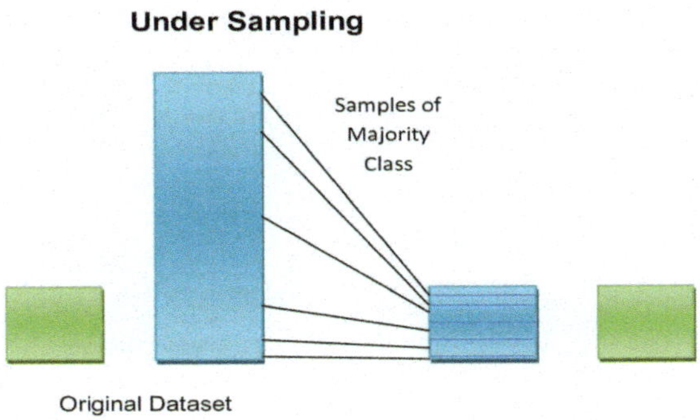

Fig. 2.4 Undersampling scenario

The sample size for both classes is now identical.

Advantages:

It requires less processing power.

Disadvantages:

- The dropping of instances may cause certain significant patterns to be lost.
- Useful only for large datasets comprising millions of instances.

Handling through algorithmic approaches: Class imbalance is naturally handled better by some algorithms than by others. Examples of these algorithms are boosting algorithms like AdaBoost and ensemble methods like random forests. Furthermore, misclassifications of instances of minority classes might be penalized more severely during training by using strategies such as cost-sensitive learning.

Custom loss function: Class imbalance during training can be lessened by creating customized loss functions that provide minority class errors greater weight or by adding class weights to regular loss functions.

Anomaly identification techniques may be more appropriate than typical classification approaches in situations where there is a severe class imbalance and the minority class reflects outliers or anomalies.

SMOTE method: An effective technique for resolving the issue of class imbalance in machine learning, particularly in deep learning applications, is synthetic minority oversampling technique (SMOTE). Whenever the minority class or classes are not adequately represented in the training dataset, it works especially well. To balance the class distribution, SMOTE creates artificial samples for the minority class [7]. The SMOTE algorithm operates in four easy steps (Fig. 2.5),

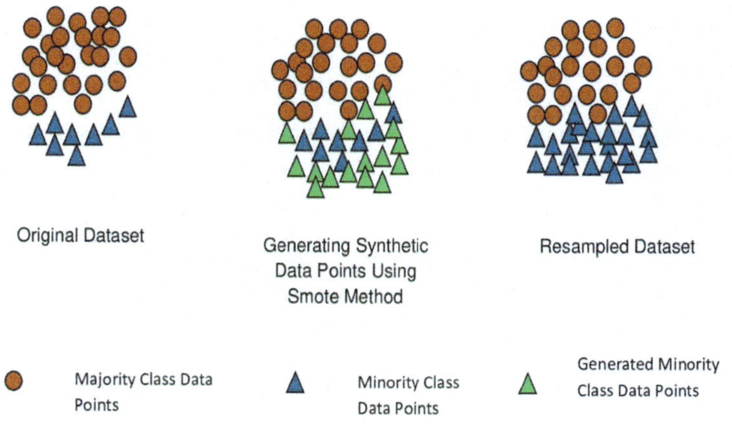

Fig. 2.5 Demonstration of SMOTE oversampling for unbalanced classification

1. Select a minority class to use as the vector input.
2. Determine its k closest neighbors (the SMOTE() function takes an argument called k_neighbors).
3. Select a neighbor from this list, and then, add a synthetic point to the line connecting the point under consideration and the neighbor of your choice.
4. Steps should be repeated until the data is balanced.

SMOTE helps address the issue of class imbalance by increasing the number of minority class instances in the training dataset without merely copying already-existing instances. SMOTE can enhance the model's capacity to generalize and produce precise predictions for instances of the minority class by generating synthetic samples that give the model more details about the minority class or classes. SMOTE also faces some challenges when overfitting may result from the generation of synthetic samples, particularly if the samples are overly similar to instances of minority classes that already exist. The number of nearest neighbors (k) in SMOTE is one parameter that can have a big impact on how effective the algorithm is. Choosing the right parameter values frequently calls for testing and verification.

2.2.5 *Sample Pseudocodes*

Oversampling

1. Load data in variable *X*. *X* <- data[, -ncol(data)]
2. Load the class of the dataset in variable *y*. *y* <- data$Class
3. Use SMOTE to oversample the minority class

   oversampled_data <- SMOTE(Class ~ ., data = data)

Undersampling

1. Find the indices of the majority class
   Majority_indices <- which(*y* == 0)
2. Sample the same number of instances from the majority class as in the minority class
   Undersampled_indices <- sample(majority_indices, sum(*y* == 1))
3. Combine the undersampled majority class with the minority class
   Undersampled_data <- rbind(data[majority_indices,], data[*y* == 1,])
4. undersampled_data now contains the undersampled dataset.

2.3 Small Disjuncts and Class Skewness

2.3.1 Small Disjuncts

Classifying a small number of training instances is known as a small disjunct. These disjuncts are of particular relevance because they are known to generate many, if not most, of the classification errors and to have a substantially greater error rate than huge disjuncts [8]. Consider a collection of animal photos in which the bulk of the images feature familiar animals such as dogs, cats, and birds. There are, however, a few pictures of uncommon or rare species as well, such as quokkas, pangolins, and aardvarks. Because they are particular, uncommon subsets that differ from the bulk of the data, these photos of unusual animals reflect little disjuncts in the dataset. In Fig. 2.6, the ● represents familiar animals, whereas ☆represents uncommon animals.

Classifier performance can be significantly impacted by small disjuncts, particularly in datasets that are imbalanced. Classifiers may be impacted by tiny disjuncts in the following ways:

Overfitting: When faced with minor disjuncts, classifiers may overfit, particularly if they are sophisticated and have few examples. Poor generalization to new data may result from this.

Misclassification: Due to the possibility that they do not have enough examples for the classifier to pick up on their distinctive characteristics, little disjuncts are more likely to be misclassified.

Fig. 2.6 Small disjuncts

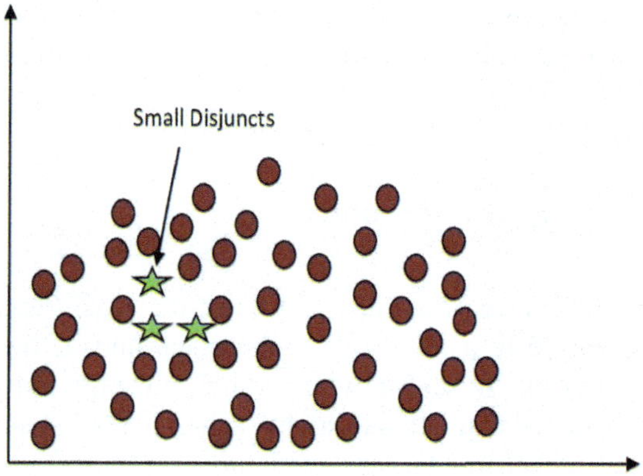

Bias: Because classifiers concentrate more on identifying the bigger, more common patterns in the data, they may be biased in favor of the majority class.

Model Performance: Minor disjuncts may have an effect on the classifier's overall performance measures, including accuracy, precision, recall, and F1-score, particularly if they are not sufficiently addressed.

2.3.2 Methods to Address Small Disjuncts

Dealing with small disjuncts is important because they can be easily overlooked or misclassified by machine learning models, leading to biased or inaccurate results [9]. There are several strategies to address the issue of small disjuncts:

- In [9] C4.5 algorithm is used because it is a well-liked learning tool for disjunctive ideas, and it could easily adapt to gather data on disjunct size statistics. C4.5 was run with its default parameters and pruning approach for the majority of the studies.
- Adaptive boosting as well as certain algorithms like cost-sensitive boosting and Ada-cost can improve the generalization of the classifier to small disjuncts by combining multiple models [10].
- For the classifier to learn from, more varied examples can be supplied by augmenting the minority class data.
- Differentiating the penalties of misclassification for distinct classes may incentivize the classifier to concentrate more on the minority class.
- The dataset can be balanced by employing strategies like undersampling (removing instances of the majority class), oversampling (replicating instances of the minority class), or more sophisticated techniques like synthetic minority oversampling technique (SMOTE), clustering-based resampling algorithm and SMOTEBoost algorithm [11].

2.3.3 Class Skewness

An imbalance in the distribution of the label (target variable) in a training dataset is referred to as skew in machine learning. If the target variable's distribution is asymmetric around its mean value, meaning that certain values are more highly represented than others and the dataset is not balanced, it is referred to as skewed. In many real-world datasets, including fraud detection, medical diagnosis, and anomaly identification, class skewness is a typical problem, where the minority class indicates uncommon events or significant outcomes.

2.3.4 *Class Skewness Types*

The three types of skewness are:

Right Skew or Positive Skew: The right side of a positively skewed distribution has a longer or fatter tail than the left. When the mean is higher than the median, the distribution is referred to be "skewed to the right." In Positive Skewness: Mean > Median > Mode.

Left Skew or Negative Skew: A distribution that is negatively skewed has a longer or fatter tail on the left side than the right. When the mean is lower than the median, the distribution is referred to as "skewed to the left." In Negative Skewness: Mean < Median < Mode

Zero Skew: When the distribution's left and right halves are roughly equal, it is said to be symmetric (zero skew). The mean and median in this instance are also approximately equal.

Fig. 2.7 shows the three types of skewness.

Class skewness has an impact on machine learning predictions' accuracy. For example, if the dataset is positively skewed, it means that it has a higher number of data points having low values. So when we train our model on this data, it will perform better at predicting the class with lower range value as compared to the class with higher values. Also, skewness tells us about the direction of outliers. If the distribution is positively skewed, then most of the outliers are present on the right side of the distribution.

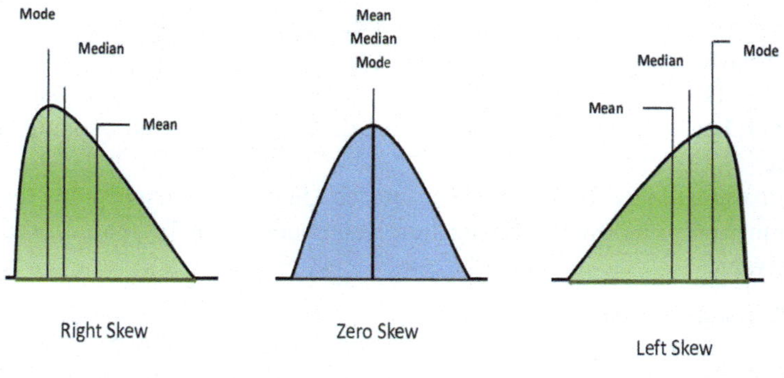

Fig. 2.7 Three types of class skewness

2.3.5 Methods to Address Class Skewness

Since the skewed data can affect our machine learning model's predicting capabilities, it is better to transform the skewed data to normally distributed data. Here are some of the ways you can transform your skewed data:

- Log transformation
- Square root transformation
- Box-Cox transformation
- Power transformation
- Sigmoid transformation.

Log Transformation

Logarithm transformation is an easy way to transform a highly skewed dataset into a more normal dataset. When modeling nonlinear variables, the probability of producing errors is also negatively skewed. Logarithm transforms the distribution of one or multiple variables to a more normal bell curve, improving the model's fit.

Square Root Transformation

Sqrt(x) is a moderate transformation that has a moderate effect on the distribution shape. It is weaker than logarithm or cube root. It is also used to reduce right skewness. It also has the advantage of being able to be applied to zero.

Box-Cox Transformation

Box-Cox is a more general type of skewness transformation. It is parameterized by a lambda value. The lambda value specifies the type of transformation to be used. Optimization techniques can help you find the optimal lambda value.

The transformation function is defined as:

$T(Y) = \left(Y^\wedge\lambda - 1\right)/\lambda$ where Y is the response variable, and λ is the transformation parameter. For $\lambda = 0$, the natural log of the data is taken instead of using the above formula.

It is parameterized by a lambda value. The lambda value specifies the type of transformation to be used. Optimization techniques can help you find the optimal lambda value.

Power Transformation

To change the distribution, raise the data to a power (such as x^2, x^3). This may work well for specific kinds of skewness.

Sigmoid Transformation

By compressing the range of values, the sigmoid function $(1/(1 + \exp^{(-x)}))$ can help lessen skewness.

It is crucial to take the analysis and data context into account while transforming skewed data. Certain distributions or types of data may lend themselves more to certain transformations. The modified data should also be examined for normality and other presumptions of the employed analysis technique.

2.4 Data Irregularities: Missing, Incomplete, Inconsistent, and Noisy Features

The real world has dirty data. Real-world datasets contain a lot of potentially erroneous data, such as instrument malfunctions, computer or human error, and transmission errors. Researchers and practitioners regularly face obstacles and features that are lacking. Incomplete or missing features can cause a number of problems and drawbacks that affect the effectiveness of machine learning models and the caliber of analysis [12]. Here are a few significant negatives:

1. Unfinished features could lead to skewed depictions of the underlying information. Inaccurate or skewed predictions can result from models trained on biased or incomplete data, which can lead to less-than-ideal decision-making.
2. Model performance might be negatively impacted by missing features. To provide accurate predictions, machine learning algorithms frequently rely on comprehensive and illuminating feature sets. Without important knowledge, models could have trouble making good generalizations to fresh data.
3. Predictions are more questionable when features are incomplete. It is difficult to predict the true underlying patterns in the data when important information is missing, which results in less consistent and more varied model outputs.
4. Predictions are more questionable when features are incomplete. It is difficult to predict the true underlying patterns in the data when important information is missing, which results in less consistent and more varied model outputs.
5. More instability may be seen in models trained on missing feature datasets. Small variations in the training data can impact on the model's behavior to become more sensitive, which reduces its robustness and dependability.
6. Imputation strategies are necessary to handle missing data, and selecting the best approach can be difficult. When dealing with missing features, imputation becomes more complex, particularly if the missingness is not random or has a relationship to other variables.
7. Absence of features results in information loss. The missing data may have important trends or insightful information that is essential for producing precise forecasts. A less thorough grasp of the underlying data distribution may arise from the failure to obtain this information.
8. Machine learning models may become more difficult to interpret if they contain incomplete features. It becomes difficult to determine each feature's contribution when some features are absent or when imputation techniques are used.

9. Computational complexity is frequently increased at the data preprocessing stage by imputing missing features. More time and processing resources might be needed for more complex imputation approaches.
10. Biases resulting from the absence of traits may have moral ramifications, particularly if they disproportionately harm particular groups. Decision-making procedures may result in unjust or discriminatory outcomes as a result.

One of the most important aspects of data preprocessing in data mining is addressing incomplete or missing features. It is imperative for researchers and practitioners to exercise caution while selecting and using imputation approaches, keeping in mind the inherent biases and uncertainties that may arise from missing data during analysis. There are different types of missing values [13].

Examples

- ***Incomplete***: It may merely contain aggregate data, lack certain relevant attributes, or have missing attribute values. Suppose, an attribute value in Fig. 2.8, *Department*=" " is missing in an employee dataset due to some human error or information not found.

  Missing data could result from,

A. Numerous tuples lack values for multiple attributes.
B. Equipment breakdown.
C. Discordant with other documented evidence, which is why it was removed.
D. Data not entered because of a miscommunication.
E. At the time of entry, certain data might not have been deemed significant.
F. Not record data modifications or histories.

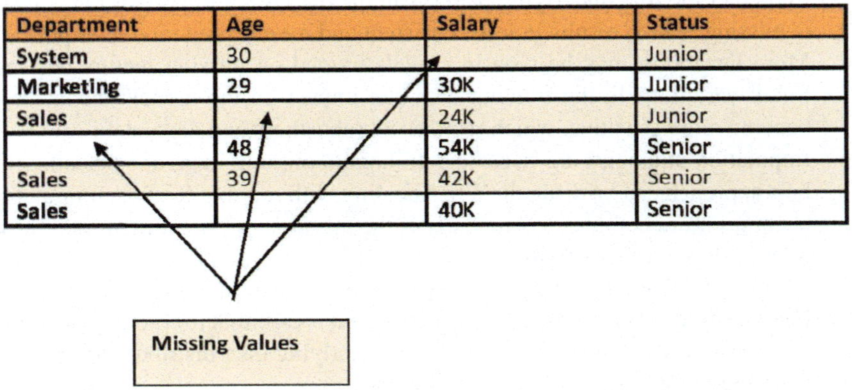

Department	Age	Salary	Status
System	30		Junior
Marketing	29	30K	Junior
Sales		24K	Junior
	48	54K	Senior
Sales	39	42K	Senior
Sales		40K	Senior

Fig. 2.8 Missing values in the employee dataset

- ***Noisy Data***: When errors or inconsistencies occur randomly or unpredictably in a dataset, it can possibly obscure the genuine underlying patterns and relationships, which is known as noisy data. A number of things can cause noise, including measurement errors, mistakes made during the data collection process, and outside influences on the data recording. It adds unwelcome variability and may cause inaccurate interpretations or forecasts when data is analyzed. A key component of data preprocessing in machine learning and data analysis is handling noisy data. If noise is ignored or handled improperly, models may not generalize well to new data or may draw incorrect conclusions. In an employee dataset, if the "*Salary= −10000*" attribute value is negative. In order to reduce the negative effects of noise on the precision and dependability of analytical insights and model predictions, robust methods for noise detection, filtering, and correction are frequently used [12, 13].

Noisy data can be two types: Class noise and attribute noise. When an example is mislabeled, class noise happens. Several factors, including subjectivity in the labeling process, mistakes in data entry, or insufficient information used to identify each sample, can be blamed for class noise.

Flaws or inconsistencies in a dataset's feature or attribute values are referred to as attribute noise. There are several possible causes of this noise, including human error, measurement flaws, and problems with data gathering. Due to the introduction of errors that the model may learn from, attribute noise can have a substantial effect on how well machine learning models perform. Figure 2.9 shows how different types of noise can impact datasets.

- ***Inconsistent***: When distinct records or variables present contradicting information inside a dataset, this is referred to as inconsistent data. Errors made when

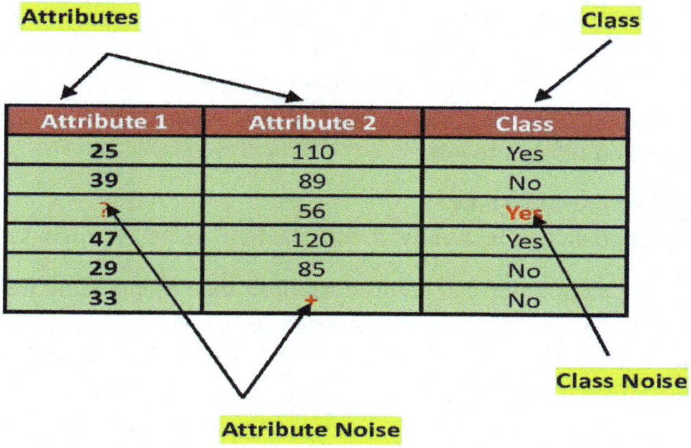

Fig. 2.9 Noise types—class noise and attribute noise

entering data, combining datasets with different formats, or irregularities in the data gathering procedure can all result in discrepancies. These discrepancies could show up as anomalies in the data structure, mismatched data types, or contradicting values. In order to guarantee data quality and dependability in analytical procedures, managing inconsistent data is essential. Data cleaning and validation approaches are frequently utilized by researchers and data practitioners to detect and address anomalies in datasets, guaranteeing that the data accurately and consistently represents the real-world phenomena being studied. In data analysis and modeling, resolving inconsistent data is essential to getting accurate and reliable findings [14].

a. There is a miscalculation between *Age*="42", *Birthday*="03/07/2010" of an employee.
b. In some portions of the dataset, the employee rating is "1, 2, 3" and now the rating is "A, B, C".
c. There may be discrepancy between duplicate records.

- ***Intentional***: Sometimes, there is a presence of disguised missing data. Those mistakes can be intentionally made, such as claiming that the birthday of every employee is 'January 1'.

There is a set of techniques by which we are able to handle all the above types of erroneous feature values. These techniques are called data preprocessing or data cleaning. The key aspect of data preparation is handling noisy, partial, and missing data in order to guarantee the correctness and dependability of machine learning models or downstream analysis [14].

1. **Handling Missing Data**

   - Use imputation techniques like mean, median, mode, or more sophisticated methods like regression imputation or k-nearest neighbors (kNN) imputation to fill in the missing data.
   - Eliminate features or instances that have missing data, particularly if they are sporadic and unrelated to the underlying patterns.
   - Fill in the missing values using some random values manually. However, it is a tedious and infeasible job.

2. **Handling Incomplete Data**
   Incomplete data can be removed with the help of the following techniques:

   - Imputation techniques can be used to fill in incomplete data, just as missing data, by predicting values based on existing information.
   - To obtain information that makes up for the incompleteness, add fresh features or alter current ones.

3. **Handling Noisy Data**
   Noisy data can be removed with the help of the following techniques:

- Binning: This method smoothes out any noise in the sorted data values by working with it. The data is separated into equal-sized buckets or bins, and each is handled separately. It is possible to substitute the mean, median, or border values for all the data in a segment.
- Use machine learning or statistical methods to find and eliminate outliers in order to lessen the influence of noisy data.
- To lessen the effect of noise on time series or continuous data, use smoothing techniques like kernel smoothing or moving through means.
- Regression: Through regression smoothing or clustering, we can detect and remove outliers.
- We can combine computer and human effort to detect suspicious values.

4. **Handling Inconsistent Data**

   There are several ways to handle inconsistent data format:

- Ascertain that the dataset's data formats are uniform. For instance, there should be a common format for dates, consistent labeling for categorical data, and the same scale for numerical values.
- Standardize the labels and categories in the dataset to a common format if they are inconsistent. Adding a single label to a group of related categories or combining semantically related categories are two examples.
- As part of the data preprocessing pipeline, include data quality tests to find and highlight inconsistent data before training the model.

Pseudocode

1. Data = load (dataset)

   # Remove duplicate data

2. Data <- drop_duplicates(inplace=True)

   # Remove Noisy data

3. dataset['feature_name'] > threshold_value for noisy_data_mask
4. dataset_withoutnoise = dataset[~noisy_data_mask]

   # Handling Missing data

5. missing_values < dataset <- [isnull().sum()]
6. imputation_method <- 'mean'
7. imputer <- SimpleImputer(strategy=imputation_method)
8. dataset_final <- DataFrame [imputer.fit_transform(dataset), columns=dataset.columns]

   # Replace inconsistent data with an appropriate value or NaN

9. inconst_data_mask <- dataset_imputed['some_column'] < some_threshold
10. dataset_cleaned <- dataset_imputed.copy()
11. dataset_cleaned.loc[inconsistent_data_mask, 'some_column'] <- np.nan

2.5 Feature Normalization

The preprocessing stage known as "feature normalization" is when we standardize the range of independent variables that make up a dataset's features. It matters because many machine learning algorithms are sensitive to the scale of the features, and various characteristics can have wildly disparate scales. Ensuring that every feature makes an equal contribution to the computation of gradients and distances during model training is the aim. This can be crucial in algorithms that depend on the size of the input features, like optimization techniques based on gradient descent [15]. Figure 2.10 demonstrates the feature normalization.

There are numerous widely used techniques for normalizing features,

A. **Min-Max Normalization**

- Sets the features' scale to a particular value, often between 0 and 1.

$$v' = \frac{v - \min_A}{\max_A - \min_A}\left(\text{new\_}\max_A - \text{new\_}\min_A\right) + \text{new\_}\min_A$$

- When you wish to bring disparate feature ranges to a single scale, this is helpful.

Ex. Let us normalize the income range from \$12,000 to \$98,000 to [0.0, 1.0]. Next, \$73,000 is assigned to $\frac{73{,}600-12{,}000}{98{,}000-12{,}000}(1.0 - 0) + 0 = 0.716$.

B. **Standardization (Z-score Normalization)**

- Sets the characteristics' mean and standard deviation to zero and one, respectively.
- $v' = \frac{v-\mu_A}{\sigma_A}$, where μ: mean, σ: standard deviation.
- This works well when the algorithm considers that the features are centered around zero, and the features have a Gaussian distribution.

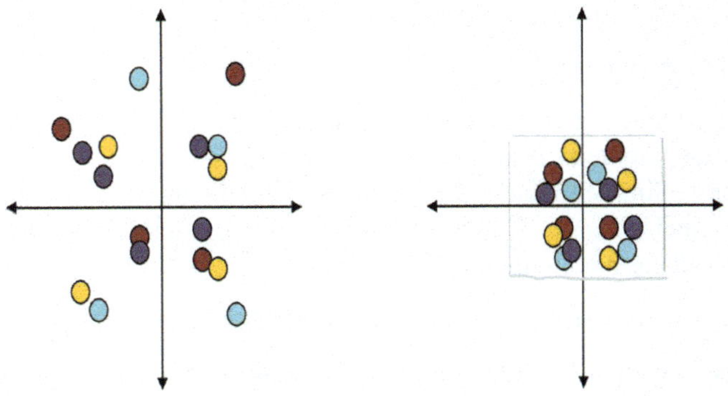

Fig. 2.10 Understanding feature normalization

Ex. Let $\mu = 54{,}000$, $\sigma = 16{,}000$. Then, $\frac{73{,}600-54{,}000}{16{,}000} = 1.225$

C. **Scaling of Decimal**

- Utilizing the median and interquartile range (IQR) rather than the mean and standard deviation is comparable to standardization.
- It works well with datasets containing extreme values because it is less susceptible to outliers.
- $v' = \frac{v}{10^j}$, where j is the smallest integer such that $\text{Max}(|v'|) < 1$.

D. **Normalization through Sum**

- Divides each feature's value by the feature's total value to scale each feature.
- Beneficial for information that shows percentages or proportions.

Example

To demonstrate feature normalization using Min-Max Scaling, let us look at a brief example. Imagine that you wish to normalize the two features in your dataset, "Age" and "Income," which are present.

Original dataset

Age	Income
25	50,000
30	75,000
35	100,000
40	60,000

Let us now normalize these features to a range of 0 to 1 by using Min-Max Scaling.

Age-Based Min-Max Scaling

$$X_{\max} = 40, X_{\min} = 25$$
$$X_{norm} = (X - X_{\min})/(X_{\max} - X_{\min})$$
$$25 \Rightarrow 0$$
$$30 \Rightarrow 0.25$$
$$35 \Rightarrow 0.5$$
$$40 \Rightarrow 1$$

Income-Based Min-Max Scaling

$$X_{\max} = 100,000, X_{\min} = 50{,}000$$
$$X_{\text{norm}} = (X - X_{\min})/(X_{\max} - X_{\min})$$
$$50{,}000 \Rightarrow 0$$

$$100{,}000 \Rightarrow 0.5$$
$$75{,}000 \Rightarrow 1$$
$$60{,}000 \Rightarrow 0.2$$

After normalization, the dataset looks like this:

Age	Income
0	0
0.25	0.5
0.5	1
1	0.2

Since "Age" and "Income" now fall into the range of [0, 1], machine learning methods that are sensitive to feature scale can use them.

The properties of your data and the needs of the machine learning algorithm you are employing will determine which normalization technique is best for you. Feature normalization is frequently used as a preprocessing step prior to model training in order to improve convergence and performance. Remember that, in order to prevent data leaking, normalization is usually applied to the training set first, and then, using the parameters (such as mean and standard deviation) computed on the training set, to the validation and test sets [16].

2.6 Overfitting and Underfitting

When a model learns the training data too well, it is prone to overfitting, a typical problem in machine learning, especially deep learning, where the fundamental patterns in the data are not captured by the model and instead are noise or random oscillations. As a result, the model works effectively with training data but not well with fresh, untested data. When a model is overly intricate in comparison with the volume and noise in the training data, this is known as overfitting. In machine learning domain, overfitting is a significant problem that must be solved using a mix of thorough evaluation methods, regularization strategies, and appropriate model construction [17].

2.6.1 Reasons of Overfitting

Complexity of Model: When a model has excessive parameters in relation to the quantity of training data, overfitting frequently occurs.

Inadequate Data: It is possible that tiny datasets do not contain sufficient different examples to allow the model to generalize well.

Data with Noise: The model may learn to fit these abnormalities rather than the true patterns if the training data contains noise or outliers.

2.6.2 Signatures of Overfitting

A machine learning model shows symptoms of overfitting when it performs remarkably well on training data but has trouble generalizing to new, unobserved data. Overfit models typically show a sharp difference in performance between training and validation, performing well on training sets with high accuracy or low loss but much worse on test or validation sets. This discrepancy indicates that the model has become too tuned to the noise or complexities in the training data rather than capturing the underlying patterns necessary for robust generalization. It is frequently accompanied by fluctuations or an increase in the validation loss following an initial decrease. For a model to be reliable and useful in real-world situations, it must be able to recognize overfitting.

2.6.3 Preventing Overfitting

Several techniques can be used to stop and lessen overfitting in machine learning models. By adding penalty terms to the model's loss function, regularization approaches like L1 or L2 regularization discourage the use of unnecessarily large weights and simplify the model. The incorporation of dropout layers, which randomly set a portion of input units to zero during training, helps keep neurons from co-adapting and improves the robustness of the model. More accurate estimates of the model's generalization performance are obtained by cross-validation approaches, which divide the dataset into training and validation sets several times. To ensure that the model learns more representative patterns, increasing the amount and diversity of the training data is another efficient way to reduce overfitting. Early stopping helps keep the model from fitting noise and guarantees that it generalizes well to new data. It stops the training process when the validation loss stops getting better. To lower the chance of overfitting, ensemble techniques like bagging and boosting combine predictions from several models. Additionally, achieving the ideal balance between underfitting and overfitting is aided by carefully tuning hyper-parameters like learning rates, batch sizes, and model architectures. For machine learning models to be more reliable and broadly applicable, a combination of these approaches that are customized to the unique properties of the dataset and issue domain must be used [18].

2.6.4 Reasons of Underfitting

In machine learning, underfitting happens when a model performs poorly on both the training set and unseen data because it is too basic to identify the hidden patterns in the training data. Underfitting is frequently caused by insufficient model complexity, which occurs when the selected technique or model architecture is unable to adequately obtain the underlying relationships in the data. The model's capacity to converge to an ideal solution may also be hampered by inadequate training time or a small number of iterations. The dataset's noisy or irrelevant features might also be a contributing issue, since they can deceive the model and keep it from discovering the real underlying patterns. Underfitting can also result from improper feature scaling or a lack of domain expertise, since the model may find it difficult to identify significant patterns in the data. Increasing model complexity, training the model for more epochs, eliminating superfluous features, enhancing feature scaling, and utilizing domain knowledge to steer the modeling process toward a more appropriate representation of the underlying data structures are common strategies used to address underfitting [19].

2.6.5 Signatures of Underfitting

When a machine learning model performs poorly on both the training and validation datasets, it is said to be underfitting since it is unable to identify the underlying patterns in the data. High training and validation errors are key markers that indicate the model is not complex enough to effectively capture the relationships in the dataset. The model has not reached its peak performance when learning curves exhibit a plateau or little improvement during training epochs. Furthermore, it is noticed that the model has poor generalization to new, unknown data, having difficulty in making accurate predictions on occurrences outside the training set. Together, these indicators suggest that the model's representation is oversimplified, and that in order to solve underfitting and improve the model's performance, changes like raising the model's complexity, training it for additional epochs, or improving feature engineering may be required.

2.6.6 Preventing Underfitting

Several techniques can be used to stop and lessen underfitting in machine learning models. Intricate patterns within the data can be captured by increasing the model's complexity by selecting more complex algorithms, using deeper structures, or adding additional layers. To enable the model to pick up on and adjust to the underlying relationships found in the training data, it must be trained for a sufficient number of epochs. A key component is feature engineering, which modifies or adds pertinent features to improve the model's capacity to identify patterns. It is imperative to maintain a suitable equilibrium between variance and bias, as an overly simplified model may be unable to adequately represent the intricacies of the data. Insights into the model's performance on various data subsets can be obtained by cross-validation, which can aid in locating and addressing underfitting. The model can be kept from oversimplifying by using regularization approaches like dropout or L1/L2 regularization. In the end, preventing and mitigating underfitting requires fine-tuning the model's architecture, optimizing hyper-parameters, and incorporating domain expertise to make sure the model accurately reflects the subtleties in the data for better generalization [18].

2.6.7 Overfitting and Underfitting Under Bias-Variance Trade-Off

- The inability of machine learning methods (like linear regression) to capture the true relationship is called bias.
- In machine learning lingo, the difference in fits between datasets is called variance.

The delicate balance between underfitting and overfitting models is addressed by the bias-variance trade-off, a basic idea in machine learning (Fig. 2.11). While variance refers to the model's sensitivity to minute fluctuations in the training data, bias refers to the inaccuracy created by oversimplifying an actual situation. There is a trade-off because decreasing bias frequently results in increasing variance and vice versa. A highly biased model performs poorly on both training and unseen data because it tends to oversimplify the underlying patterns. Conversely, a model with high variance does well when training but poorly when applied to fresh data since it picks up noise in the training set. Finding the best balance between minimizing bias and variance is the aim of machine learning, which aims to produce a model that captures the key patterns in the training set and has good generalization to new data. To create machine learning models that are reliable and accurate, this careful trade-off is essential [20].

The phenomena of overfitting and underfitting in machine learning are significantly influenced by the bias-variance trade-off. When a model is overly intricate, it fails to generalize successfully to new, unobserved data, capturing noise and specificities in the training set. This phenomenon is known as overfitting. This is frequently linked to high variance and low bias. On the other side, underfitting arises when a model lacks complexity and struggles to discern the fundamental patterns within the data, leading to unsatisfactory performance on both training and test sets. Usually, underfitting is linked to low variance and strong bias. In order to reduce overfitting and underfitting, the bias-variance trade-off must be taken into consideration. Regularization techniques can be used to penalize overly complicated models, preventing overfitting and reducing variance, if the model is too complex. On the other hand,

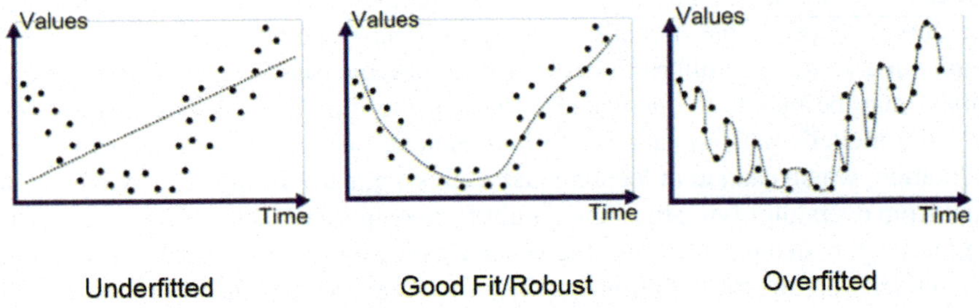

Fig. 2.11 Sample plots of overfitting, underfitting, and best fit situations

if the model is overly simplistic, employing more complex techniques or making it more complex can assist prevent underfitting and reduce bias. In the end, striking the correct balance between variance and bias is crucial for developing machine learning models that perform consistently across a variety of datasets and scenarios and that generalize well to new data. In order to traverse the bias-variance trade-off and achieve this delicate balance in machine learning, regularization algorithms, cross-validation, and cautious model selection are often used methodologies [20].

Example

Suppose, we have two datasets of 10,000 instances and 1,000,000 instances, respectively, and they are divided into training, development, and testing set accordingly.

Dataset-1 (10,000)	Training set—70%	Development set—15%	Test set—15%
	Training set—60%	Development set—20%	Test set—20%
Dataset-2 (1,000,000)	Training set—98%	Development set—1%	Test set—1%
	Training set—99.5%	Development set—0.4%	Test set—0.1%

Train set error	1%	15%	15%	0.5%
Development set error	11%	16%	30%	1%
Bias-variance trade-off	High variance (overfitting)	High bias (underfitting)	High bias and high variance	Low bias and low variance

2.7 Impact of Class Imbalance and Data Irregularities in Multiclass, Multi-objective, and Multilabel Classification

Machine learning models can be seriously hindered in terms of performance and reliability by class imbalance and irregularities in the data when it comes to multiclass categorization. In particular, for minority classes, imbalance skews model training toward the majority class, producing biased predictions and inadequate generalization. Evaluation metrics become deceptive when they prioritize accuracy over minority class performance. Furthermore, these difficulties are made worse by data abnormalities like missing values, outliers, or incorrectly labeled cases, which distort decision boundaries and raise the number of false positives and negatives. To ensure that the model is useful for all classes, mitigating these effects calls for meticulous preprocessing, specialized sample procedures, and strong assessment mechanisms. Class imbalance and irregular data can have a significant impact on the model's capacity to prioritize and balance numerous objectives in multi-objective classification. The difficulty of properly valuing objectives is made more difficult by class imbalance, which may result in biased model training and the priority of dominant objectives over others. The presence of noise or inconsistencies in the data might cause problems for the model, making it more difficult for it to identify significant patterns across a variety of objectives. This may lead to less-than-ideal choices and impaired performance, especially when goals clash or vie for attention. To tackle these problems, great thought must be given to objective weighting, strong data preparation methods, and sophisticated model architectures that can support several, sometimes-incompatible aims while preserving efficacy and fairness. Class imbalance and irregular data can have a major impact on the model's capacity to predict several labels for each instance in multi-label classification. An imbalance in classes could result in biased training, which would make the model prefer labels that appear frequently to minority labels. This would mean that less represented labels would perform worse. Furthermore, anomalies in the data, like as noise, missing values, or cases with incorrect labels, might interfere with the learning process and make it difficult for the model to identify significant patterns across a range of labels. Reduced prediction accuracy and dependability may result from this, especially in cases involving uncommon or irregular label combinations. Tackling these obstacles calls for specialized preprocessing methods, strong class imbalance management, and approaches to lessen the effect of data anomalies so that the model can effectively predict numerous labels for every occurrence.

2.8 Conclusions

It is critical to manage class imbalance and irregular data in classification tasks in order to guarantee the accuracy and resilience of machine learning models. By utilizing several preprocessing methods including data cleaning, normalization, and feature engineering, we can lessen the influence of anomalies like noise, outliers, and missing values on the performance of the model. Furthermore, as class imbalance might result in biased classifiers that underperform on minority classes, it is important to carefully address the issues it poses. Future research should concentrate on creating new preprocessing techniques and improving current ones in order to better manage class imbalance and irregular data. Furthermore, in order to advance the discipline and develop classifiers that perform well across a variety of datasets, it will be imperative to investigate alternate evaluation metrics and methodologies for model evaluation in imbalanced circumstances. In the end, we can enhance the accuracy and usefulness of categorization models in practical situations across a range of domains by tackling these issues.

Bibliography

1. Das S, Datta S, Chaudhuri BB (2018) Handling data irregularities in classification: foundations, trends, and future challenges. Pattern Recogn 81:674–693
2. Burez J, Van den Poel D (2009) Handling class imbalance in customer churn prediction. Expert Syst Appl 36(3):4626–4636
3. Zhang L, Yang H, Jiang Z (2018) Imbalanced biomedical data classification using self-adaptive multilayer ELM combined with dynamic GAN. Biomed Eng Online 17(1):1–21
4. Wang S, Minku LL, Yao X (2014) Resampling-based ensemble methods for online class imbalance learning. IEEE Trans Knowl Data Eng 27(5):1356–1368
5. Sharma S, Gosain A, Jain S (2022) A review of the oversampling techniques in class imbalance problem. In: International conference on innovative computing and communications: proceedings of ICICC 2021, vol 1. Springer, pp 459–472
6. Peng M, Zhang Q, Xing X, Gui T, Huang X, Jiang YG et al (2019) Trainable undersampling for class-imbalance learning. In: Proceedings of the AAAI conference on artificial intelligence, vol 33, No. 01, pp 4707–4714
7. Guo H, Zhou J, Wu CA (2018) Imbalanced learning based on data-partition and SMOTE. Information 9(9):238
8. Weiss GM (2009) The impact of small disjuncts on classifier learning. Data Mining: special issue in annals of information systems. Springer, Boston, pp 193–226
9. Weiss GM (1994) The problem with noise and small disjuncts. Rutgers University
10. Freund Y, Schapire RE (1995) A decision-theoretic generalization of on-line learning and an application to boosting. In: Lecture notes in computer science (including subseries lecture notes in artificial intelligence and lecture notes in bioinformatics), vol 904, pp 23–37
11. Chawla NV, Lazarevic A, Hall LO, Bowyer KW (2003) "SMOTEBoost: improving prediction of the minority class in boosting. In: Knowledge discovery in databases: PKDD 2003: 7th European conference on principles and practice of knowledge discovery in databases, Cavtat-Dubrovnik, Croatia, vol. 2838, 22–26 Sept 2003. Springer, Berlin, pp 107–119
12. Somasundaram RS, Nedunchezhian R (2011) Evaluation of three simple imputation methods for enhancing preprocessing of data with missing values. Int J Comput Appl 21(10):14–19

13. Alasadi SA, Bhaya WS (2017) Review of data preprocessing techniques in data mining. J Eng Appl Sci 12(16):4102–4107
14. Jansen BJ, Aldous KK, Salminen J, Almerekhi H, Jung SG (2023) Data preprocessing. In Understanding audiences, customers, and users via analytics: an introduction to the employment of web, social, and other types of digital people data. Springer, Cham, pp 65–75
15. Singh D, Singh B (2020) Investigating the impact of data normalization on classification performance. Appl Soft Comput 97:105524
16. Ahmad T, Aziz MN (2019) Data preprocessing and feature selection for machine learning intrusion detection systems. ICIC Express Lett 13(2):93–101
17. Montesinos López OA, Montesinos López A, Crossa J (2022) Overfitting, model tuning, and evaluation of prediction performance. Multivariate statistical machine learning methods for genomic prediction. Springer, Cham, pp 109–139
18. Jabbar H, Khan RZ (2015) Methods to avoid overfitting and under-fitting in supervised machine learning (comparative study). Comput Sci Commun Instrum Dev 70(10.3850):978–981
19. Cunningham P, Delany SJ (2021) Underestimation bias and underfitting in machine learning. In: Trustworthy AI-integrating learning, optimization and reasoning: first international workshop, TAILOR 2020, Virtual Event, 4–5 Sept 2020, Revised Selected Papers 1. Springer, Berlin, pp 20–31
20. Briscoe E, Feldman J (2011) Conceptual complexity and the bias/variance tradeoff. Cognition 118(1):2–16

Chapter 3
Multi-class Classification

3.1 Introduction

In recent years, there has been a significant growth in the number and quality of classification models constructed from data. The number of labeled data examples available to train different machine learning models is an important determinant of their progress and improvement. In the supervised machine learning classification process, the model attempts to predict the correct label of a given input set of data. Before being utilized to make predictions on fresh, unobserved data, the model in classification is thoroughly trained to utilize the training set and then assessed using test data. Classification can be divided into two types: (a) binary and (b) multi-classification. Now, binary classification categorizes data into one of two classes or categories. It has a variety of applications, including spam email identification, medical diagnosis, sentiment analysis, fraud detection, and many others. On the other hand, multi-class classification involves assigning a new instance to at least three classes. When there are more classes in a classification problem, correctly categorizing a new instance becomes more difficult compared to making the same judgment with fewer classes.

For example, binary classification for heart disease detection entails developing a model that predicts whether a person has heart disease (positive class) or not (negative class) based on a variety of input features (Fig. 3.1a) [1]. The main goal is to find the class or category that the new data belongs to. Multi-class classification for heart disease detection involves predicting multiple classes, e.g., 1–4 (low, medium, high, and serious), each number representing a different stage or type of heart disease (Fig. 3.1b).

S. Chakraborty and L. Dey, *Multi-objective, Multi-class and Multi-label Data Classification with Class Imbalance*, Springer Tracts in Nature-Inspired Computing,
https://doi.org/10.1007/978-981-97-9622-9_3

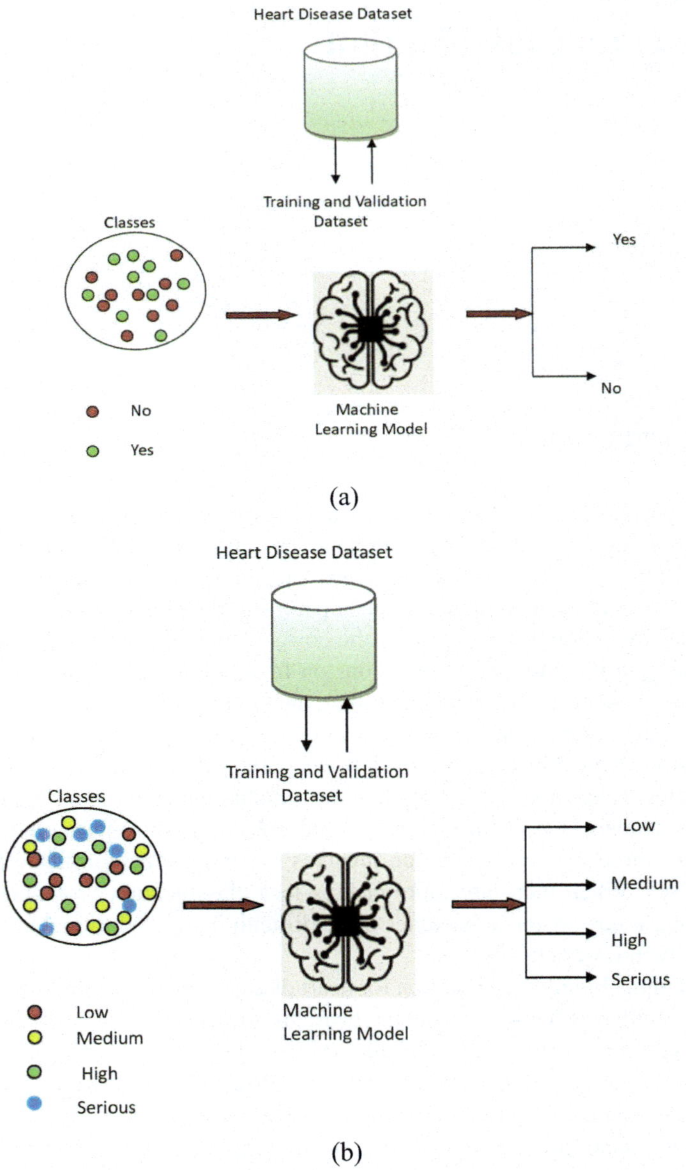

Fig. 3.1 Heart disease prediction using **a** binary classifier, **b** multi-class classifier

3.2 Binary Classification

In conventional classification tasks, input samples equate to a single target label. This sort of classification is known as single-label classification [2]. Binary categorization divides incoming data samples into two categories depending on a given metric. Binary classification is the fundamental condition for any classification approach.

All known classification methods and modifications can be applied to binary classification. The normal situation is represented by one class in most binary classification problems, and the aberrant condition is represented by the other [3]. It may be necessary to forecast if a potential consumer will make a purchase, whether the mail is spam or not, whether a medical test reveals the presence of an illness, or whether it rained that day (1 for rain, 0 for no rain). Every one of these classification tasks is binary. Random forests, Naive Bayes, K-nearest neighbor, logistic regression, support vector machines, and decision trees are examples of common techniques for binary classification. The training data used by these algorithms is labeled, with each example designated as belonging to one of the two groups. After that, the model makes predictions on fresh, unobserved data using the training set.

3.2.1 Classification Performance Assessment

Assessing the performance of a classification model is summarized by a table known as a contingency table or confusion matrix. A typical confusion matrix template for a binary classifier is divided into four quadrants (Fig. 3.2): The top-left box displays the number of true positives (TP), which represents the number of valid predictions for the positive class. The box beneath it contains false positives (FP), which are occurrences of the negative class that were wrongly recognized as positive. These are known as type I errors in statistics. The top-right box shows the number of false negatives (FN), which are actual positive events that were incorrectly projected as negatives. Finally, the bottom-right box shows the number of true negatives (TN), which are genuine negative class instances that are correctly predicted as negative.

Let us consider, a heart disease dataset had two labels—Yes and No (Table 3.1). Each class has 40 instances. The first row states that out of 40 positives, 31 are correctly predicted and 9 incorrectly (Marked in red). Similarly, the second row highlights that out of 40 negatives, 12 (marked in red) are incorrectly predicted and 28 correctly.

Three popular measures that separately estimate a classifier's performances on different classes are accuracy, precision, and recall.

Accuracy

Accuracy is the frequency with which the classifier makes valid predictions. We can define accuracy as the ratio of the number of correct forecasts to the total number of predictions. Accuracy score of the binary classifier is as follows,

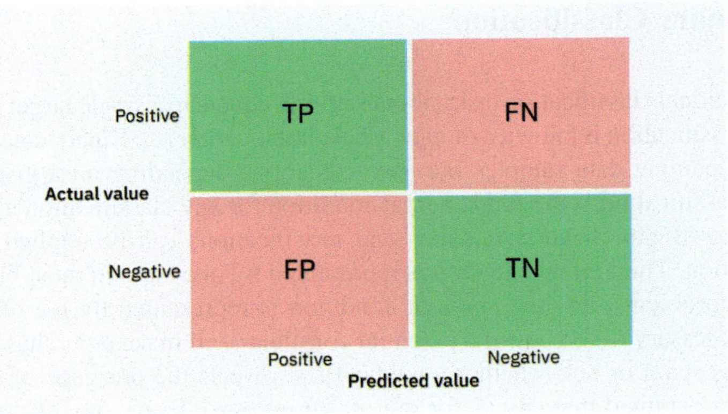

Fig. 3.2 Confusion matrix template

Table. 3.1 Two-class confusion matrix for assessing classifier performance

	Predicted (yes)	Predicted (no)	Total
Actual (yes)	31	9	40
Actual (no)	12	28	40
Total	43	37	80

$$\text{Accuracy} = \frac{\text{TP} + \text{TN}}{\text{TP} + \text{TN} + \text{FP} + \text{FN}}$$

From Table 3.1, the accuracy can be calculated as $= ((31 + 28)/31 + 28 + 12 + 9) * 100 = 73.75\%$.

Precision (Positive Predictive Value)

Precision is defined as the ratio of accurately predicted positive observations to all expected positives. It evaluates the accuracy of optimistic predictions,

$$\text{Precision} = \frac{\text{TP}}{\text{TP} + \text{FP}}$$

From Table 3.1, the precision can be calculated as $= (31/31 + 12) * 100 = 72.09\%$.

Recall

Recall is the ratio of accurately predicted positive observations to all observations in the class. It tests the model's ability to find all of the relevant cases,

$$\text{Recall} = \frac{\text{TP}}{\text{TP} + \text{FN}}$$

From Table 3.1, the recall can be calculated as $= (31/31 + 9) * 100 = 77.5\%$.

- **A Python example for binary classification**

Step 1: Load the Libraries

First, import the necessary libraries.

```
import pandas as pd
from sklearn import metrics
import matplotlib.pyplot as plt
from sklearn.model_selection import train_test_split
From sklearn.metrics import classification_report, confusion_matrix, accuracy_score
```

Step 2: Load the Data and Fit a Classifier Model to the Training Data

We use 80% of the data for training and 20% for testing. Train a logistic regression model on the training data and then make predictions on the testing data.

```
from sklearn.linear_model import LogisticRegression
df = pd.read_csv('data.csv')
X = df['data']
y = df['target']
X_train, X_test, y_train, y_test = train_test_split(X, y, test_size=0.2, random_state=42)
model = LogisticRegression()
model.fit(X_train, y_train)
predictions = model.predict(X_test)
```

Step 3: Calculate the Accuracy Score by Comparing the Actual Values and Predicted Values.

We can now determine how well the model performed by comparing its predictions to the actual target values, which we saved in the *y*_test variable. First, we will compute the confusion matrix to get the necessary parameters:

```
mat=confusion_matrix(actual, predicted)
print("Confusion Matrix:\n", t)
cm_display = metrics.ConfusionMatrixDisplay(confusion_matrix = mat, display_labels = [0,
1])
cm_display.plot()
plt.show()
```

With these values, we can now calculate an accuracy, precision and recall.

```
Accuracy = metrics.accuracy_score(actual, predicted)
print("Accuracy=", Accuracy)
Precision = metrics.precision_score(actual, predicted)
print("Precision=",Precision)
Sensitivity_recall = metrics.recall_score(actual, predicted)
print("Recall=", Recall)
```

In Fig. 3.3, the confusion matrix shows the actual and predicted labels.

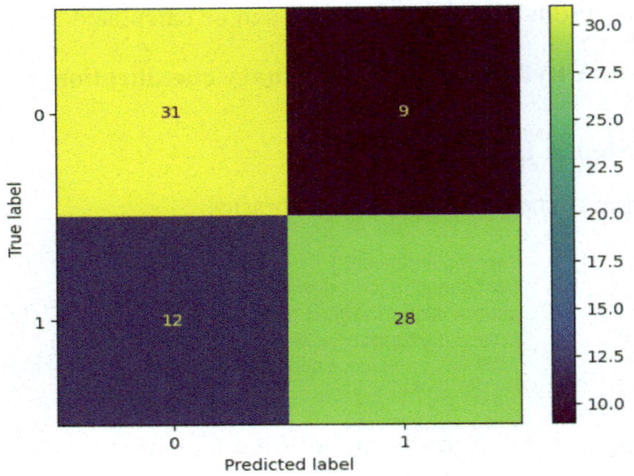

Fig. 3.3 Two-class confusion matrix for the example of Table 3.1

3.2.2 *Binary Classification Pitfalls*

Typically, a binary classification model assigns a class based on a threshold after producing a probability score ranging from 0 to 1. For instance, the input is categorized as positive if the score is higher than 0.5 and as negative otherwise. While some classification algorithms work well for any kind of classification problem, not all techniques can be used for binary classification situations. There are several pitfalls of binary classification,

(a) Handling more than two classes: Problems where there are only two classes or categories to predict are specifically intended for binary classification. However, if there are more than two target classes, we have to go for multi-class classification. Let us now look at a dataset including pictures of handwritten numbers (0–9). Our task is to categorize each picture into one of the ten possible digits. Since each image can be assigned to one of the following 10 classes: 0, 1, 2, 3, 4, 5, 6, 7, 8, or 9, this is a multi-class classification problem. A multi-class classifier would be utilized to forecast the digit that each picture corresponds to.

(b) Hidden subclass: In binary classification, the hidden subclass problem arises when data in one of the classes contains several unique but unseen subgroups. This may result in some problems, including poorer model performance and false conclusions. The model might not perform well because it ignores the unique patterns within subclasses and treats all instances within a class as similar.

(c) Problems with nonlinear decision boundaries: Logistic regression and other binary classifiers are based on the assumption of linear decision boundaries. The complexity of the solution may increase in situations when the decision boundary is nonlinear, and more sophisticated models, such as SVMs or neural networks, may be needed.

3.3 Multi-class Classification

Until now, only binary categorization has been considered. In many machine learning applications, more than two classes may be used. For instance, there are 26 classes for handwritten alphabet recognition. Therefore, this classification cannot be solved with a binary one. A classification task with more than two classes is known as multi-class classification, which is essential for a wide range of real-world machine learning applications that require the ability to distinguish between thousands of different classes automatically [4]. Multi-class classification challenges emerge in various disciplines, including biology, computer vision, social network research, and information retrieval, where data examples may belong to multiple classes.

An instance that frequently calls for multi-class categorization is categorizing a collection of pictures of fruit having four classes, such as bananas, apples, oranges, and grapes (Fig. 3.4).

In multi-class categorization, we must assign instances to one of three or more classes [5]. There are three major types of approaches used to tackle multi-class classification problems,

(a) Methods extended from binary case: Several machine learning algorithms can be expanded from binary to multi-class classification. For example, decision trees and random forests may naturally manage many classes by separating nodes depending on feature discriminability and mixing multiple trees to improve accuracy. Neural networks employ a softmax output layer to simulate complex interactions. Support vector machines (SVM) with extensions such as the Crammer–Singer formulation for dealing with multiple classes and gradient boosting machines (GBM), which construct models sequentially to repair earlier errors and increase overall prediction accuracy.
(b) Hierarchical classification: Hierarchical classification for multi-class problems arranges classes in a tree-like structure, with judgments made at different levels of the hierarchy, gradually decreasing the classification. This method uses domain knowledge or class similarities to form the hierarchy, and predictions

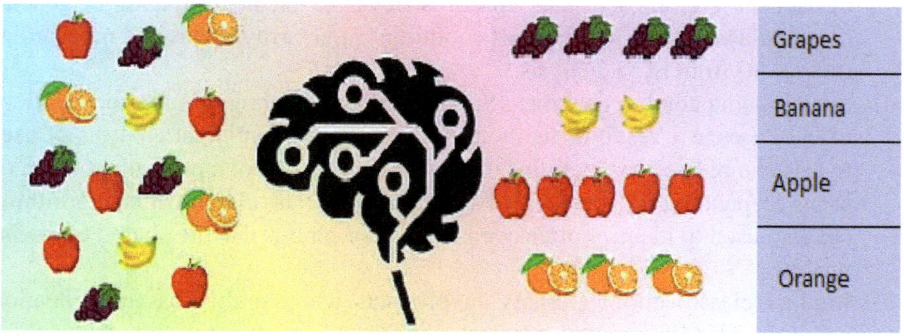

Fig. 3.4 Multi-class classification in machine learning

are made by traversing the tree from root to leaf node. This method simplifies complex classification tasks, improves accuracy by dividing down decisions into simpler sub-problems, and reduces computer cost as compared to flat classification approaches.

(c) Methods converting multi-class into binary classification problems: Methods for converting multi-class problems into binary classification problems include one-versus-all (OvA), where a separate binary classifier is trained for each class against all other classes, and the class with the highest confidence score is chosen; one-versus-one (OvO), which involves training a binary classifier for every possible pair of classes and using a voting scheme to determine the final class; and error-correcting output codes (ECOC), which creates a coding matrix. These strategies reduce the multi-class problem to more manageable binary tasks, which improves flexibility and, in some cases, accuracy.

3.3.1 Applications of Multi-class Classification

Sorting instances into one or more classes is known as multi-class classification. It is widely applicable in many different fields. Here are a few typical and noteworthy uses:

(a) Malware categorization: An important application of multi-class classification is malware categorization. Because malware attacks and threats are constantly changing due to the rapid advancements in technology, more sophisticated algorithms are being utilized to determine if a certain piece of software is malware or not.

(b) Image classification: It is a common application of classification in which various classes can be applied to an image. For instance, we can categorize a person's emotion in a certain image as being happy, shocked, surprised, angry, etc. Multi-class picture classification also includes face recognition.

(c) Medical diagnosis: Determining the presence and severity of a certain disease in a patient is a multi-class problem. A prevalent use case is identifying the type of cancer (e.g., breast, lung, and prostate) from medical imaging or genetic data or classifying different heart conditions (e.g., arrhythmias and myocardial infarction) from ECG analysis.

(d) Social media content analytics: Social media is a data goldmine for a variety of use cases as a result of its increased popularity. Sentiment analysis of user tweets can be used to determine the general perception of a particular person or thing. Depending on their level of severity, multi-class classification algorithms are also used to identify offensive language or photos that are shared on social media.

(e) Product classification: Classifying products using multi-class classification entails dividing them into several preestablished groups according to their characteristics. For e-commerce platforms to classify things like electronics, apparel, and home goods, this is crucial. Businesses can automate the categorization of new inventory, increasing search and recommendation accuracy, by training models on labeled data. For this, methods like random forests, decision trees, and neural networks are frequently employed.

3.4 Types of Multi-class Classification

A multi-class classification is a classification approach that allows to categorization of test data into multiple class labels seen in trained data to forecast a model. There are two sorts of multi-class classification techniques: one-versus-all (one-vs-rest) and one-versus-one.

3.4.1 One-Versus-All (One-vs-Rest) (OvR)

In one-versus-rest (OvR) classification, we must create N-binary classifier models for the dataset of N-class occurrences. The number of class labels in the dataset and the number of created binary classifiers must be equal. It requires breaking down the multi-class dataset into discrete binary classification problems. On each binary classification problem, a binary classifier is trained, and the model with the highest confidence is used to create predictions.

For example, consider a multi-class classification problem with instances for each of the three classes: red, blue, and green. This might be separated into three binary classification datasets, as follows.

Classifier 1: [Green] versus [Red, Blue]
Classifier 2: [Blue] versus [Green, Red]
Classifier 3: [Red] versus [Blue, Green].

A pictorial representation of one-versus-all multi-class classification is depicted in Fig. 3.5.

3.4.1.1 Advantages of One-Versus-All Classification

One-versus-all (OvA) classification, also known as one-against-all, is a prominent technique for dealing with multi-class classification problems. Here are some of the primary benefits of this approach:

(1) **Simplicity**: OvA is easy to apply. It entails developing a separate binary classifier for each class, with each classifier distinguishing between one and all other classes.
(2) **Scalability**: OvA works well with problems involving a large number of classes. Because each binary classifier only needs to discriminate between one class and

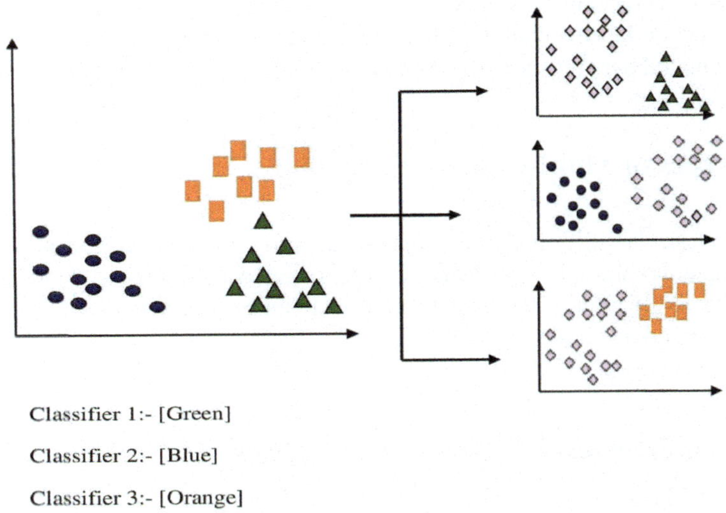

Fig. 3.5 One-versus-rest (OvR) multi-class classification

the others, its complexity is minimized when compared to methods that handle all classes concurrently.

(3) **Flexibility**: Each classifier can be trained using any binary classification algorithm. This implies you can select the optimal binary classifier for your particular situation, such as logistic regression, support vector machines, decision trees, or others.

(4) **Ease of Training**: Training individual binary classifiers may be more computationally feasible than training a single multi-class classifier, particularly when using methods that do not inherently allow multi-class classification.

(5) **Robustness**: If one classifier fails or performs poorly, it does not necessarily affect the performance of the other classifiers. This makes the OvA technique more resilient than approaches that consider all classes concurrently.

3.4.1.2 Limitations of One-Versus-All Classification

While the one-versus-all (OvA) classification has many advantages, it also comes with several limitations that can affect its performance and applicability in certain scenarios. Here are some of the key limitations:

(1) **Imbalance**: The technique may cause class imbalance concerns, particularly if the classes in the dataset are uneven. This can have an impact on how well individual classifiers perform.

(2) **Computational Complexity**: Training N classes can be computationally costly, particularly for big datasets and multiple classes. This can result in longer training times and increased computational resource requirements.

(3) **Dependence on Binary Classifiers**: The performance of one-versus-all classification is highly influenced by the performance of the underlying binary classifiers. If the binary classifiers do not perform well, the multi-class classifier's overall performance suffers.
(4) **Inconsistent Predictions**: Inconsistencies in predictions may occur when numerous classifiers produce high confidence scores for different classes. This may necessitate additional rules or tie-breaking procedures to settle issues, complicating the decision-making process.
(5) **Class Overlap**: When classes are not segregated, binary classifiers may struggle to discriminate between positive and negative classes. This can result in poor classification performance, particularly where class borders are unclear.

3.4.2 One-Versus-One (OvO)

The method involves training a binary classifier for each pair of classes—that is, one class against all other classes. The following straightforward formula can be used to determine the number of binary classifiers that need to be trained:

$$(N * (N - 1))/2 \quad \text{where } N = \text{total number of classes.}$$

For example, consider a classification problem with three sorts of classes: green, blue, and orange.

We partition this problem into $N * (N - 1)/2 = 3$ binary classifier tasks (Fig. 3.6).

Classifier 1: Orange versus Green
Classifier 2: Blue versus Green
Classifier 3: Orange versus Blue

For each object, probabilities for each class are summed up, and the class with the largest probability score is used for classification. The final class prediction is determined by taking the majority of the predictions from all binary classifiers, each of which predicts a single class. This approach is most effective when applied to kernel-based algorithms such as SVM. A pictorial representation of one-versus-all multi-class classification is depicted in Fig. 3.6.

3.4.2.1 Advantages of One-Versus-One (OvO) Classification

One-versus-one (OvO) classification, also known as one-against-one, is another method for dealing with multi-class classification problems [6]. It entails training a unique classifier for each pair of classes. Here are the primary benefits of this approach:

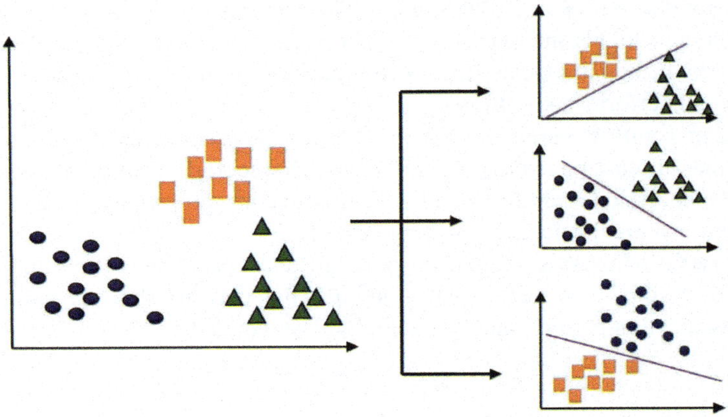

Fig. 3.6 One-versus-one (OvO) multi-class classification

(1) Reduction in Class Imbalance: Each binary classifier is trained on a pair of classes, which produces more balanced datasets than one-versus-all (OvA). This can lead to more accurate and less biased classifiers.
(2) Simpler Decision Boundaries: Each binary classifier just has to distinguish between two classes, resulting in simplified decision limits. This simplicity has the potential to increase individual classifier performance while also strengthening the entire model.
(3) Scalability to Many Classes: Although the number of classifiers increases quadratically with the number of classes, each classifier is trained on a smaller subset of the data. This can make training more manageable and economical, particularly when using parallel computing resources.
(4) Voting-Based Decision Making: The final class is usually selected by a majority vote among all paired classifiers. This can result in more consistent and dependable predictions.

3.4.2.2 Limitations of One-Versus-One (OvO) Classification

(1) Quadratic Growth of Classifiers: The number of binary classifiers needed increases quadratically with the number of classes. This can result in significant increases in computational and memory needs, particularly for a large number of classes. Also managing a large number of classifiers might complicate the implementation.
(2) Challenges with Voting Mechanisms: Determining the final class label based on the votes of many classifiers can be difficult, especially when votes are tied or close. This may necessitate additional decision rules or tie-breaking procedures, complicating the prediction process.

(3) Overfitting: Many classifiers run the danger of overfitting, especially if some classes are underrepresented in the training data. Overfitting can cause poor generalization to new, unseen data, lowering the model's overall performance.

3.4.3 Case Study on One-Versus-One, One-Versus-Rest with SVM

In multi-class classification circumstances, we must deal with more than two classes, and hence, the algorithm we use must be able to handle multiple classes. There are various classic models for this. For example, decision tree classifier, random forest classifier,·XGBoost, Naïve Bayes, support vector machines (SVM), K-nearest neighbor (KNN) AdaBoost, etc., can handle multi-class classification. In this section, we will apply SVM for multi-class classification on the iris dataset. Both one-versus-rest (OvR) and one-versus-one (OvO) approaches primarily break multi-class data into binary classification data, which can then be converted using binary classification algorithms [7].

The iris dataset includes four features: petal length, petal width, sepal length, and sepal width from 50 samples of three Iris species (Iris setosa, Iris virginica, and Iris versicolor) (Figs. 3.7 and 3.8). These measurements were used to build a linear discriminant model to categorize the species. Iris datasets are frequently used as benchmarks to evaluate classification techniques like support vector machines, decision trees, and k-nearest neighbors. They are commonly used in beginner machine learning courses, research articles, and tutorials due to their simplicity and applicability.

Fig. 3.7 Iris flower species (Taken from Machine Learning in R for beginners) (https://www.datacamp.com/tutorial/machine-learning-in-r)

	sepal_length	sepal_width	petal_length	petal_width	species
0	5.1	3.5	1.4	0.2	Iris-setosa
1	4.9	3.0	1.4	0.2	Iris-setosa
2	4.7	3.2	1.3	0.2	Iris-setosa
3	4.6	3.1	1.5	0.2	Iris-setosa
4	5.0	3.6	1.4	0.2	Iris-setosa
...	...	...	...	...	...
145	6.7	3.0	5.2	2.3	Iris-virginica
146	6.3	2.5	5.0	1.9	Iris-virginica
147	6.5	3.0	5.2	2.0	Iris-virginica
148	6.2	3.4	5.4	2.3	Iris-virginica
149	5.9	3.0	5.1	1.8	Iris-virginica

150 rows × 5 columns

Fig. 3.8 Iris dataset summary

The dataset contains three classes: Setosa, versicolor, and virginica. Therefore, it can be perfectly used to demonstrate the multi-class classification using SVM.

One-Versus-Rest (OvR)

The one-versus-rest (OvR) or one-versus-all (OvA) approach divides the multi-class data into binary classification data, which may then be converted using binary classification algorithms. The translated binary classification data will appear as follows,

Setosa against [Versicolor, Virginica]
Versicolor against [Setosa, Virginica]
Virginica against [Setosa, Versicolor].

Here is an example of how we can use the iris dataset for multi-class classification using Python and the scikit-learn library. Scikit-learn libraries include a method named OneVsRestClassifier in the multi-class package that can be used to implement the one-versus-rest (OvR) technique.

1. **Import Libraries First**

```
from sklearn import datasets
from sklearn.svm import SVC
from sklearn.multiclass import OneVsRestClassifier
import matplotlib.pyplot as plt
from sklearn import metrics
from sklearn.metrics import classification_report, confusion_matrix, accuracy_score
```

2. **Load the Iris Dataset and Apply the One-Versus-Rest SVM for Classification**

```
iris =datasets.load_iris()

X=iris.data

y =iris.target

svc =SVC()

Train the model

OvR=OneVsRestClassifier(svc)

OvR.fit(X,y)

Predicting values from the model.

Result=OvR.predict(X)

Result
```

Output

```
array([0, 0, 0, 0, 0, 0, 0, 0, 0, 0, 0, 0, 0, 0, 0, 0, 0, 0, 0, 0, 0, 0,
       0, 0, 0, 0, 0, 0, 0, 0, 0, 0, 0, 0, 0, 0, 0, 0, 0, 0, 0, 0, 0, 0,
       0, 0, 0, 0, 0, 0, 1, 1, 1, 1, 1, 1, 1, 1, 1, 1, 1, 1, 1, 1, 1, 1,
       1, 1, 1, 1, 1, 1, 1, 1, 1, 1, 1, 2, 1, 1, 1, 1, 1, 2, 1, 1, 1, 1,
       1, 1, 1, 1, 1, 1, 1, 1, 1, 1, 1, 1, 2, 2, 2, 2, 2, 2, 1, 2, 2, 2,
       2, 2, 2, 2, 2, 2, 2, 2, 2, 1, 2, 2, 2, 2, 2, 2, 1, 2, 2, 2, 2, 2,
       2, 1, 2, 2, 2, 2, 1, 2, 2, 2, 2, 2, 2, 2, 2, 2, 2, 2])
```

3. **Print the Confusion Matrix**

```
matl=confusion_matrix(y,Result)

print("Confusion Matrix:\n",t)

cmatl_display =metrics.ConfusionMatrixDisplay(confusion_matrix =matl,display_labels=[0,
1,2])

cmatl_display.plot()

plt.show ()
```

Output

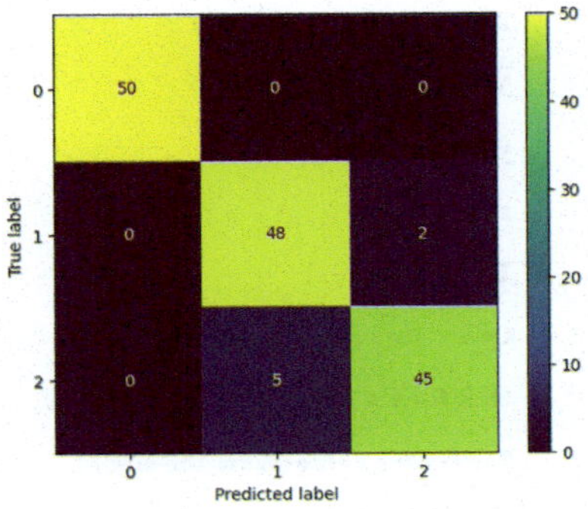

One-Versus-One (OvO)

This strategy involves splitting data for each class, with each class facing every other class as an opponent. We may use the iris dataset again to better understand how this strategy splits data. The iris contains the following classes:

Setosa
Versicolor
Virginica.

Splitting iris data using the one-versus-one (OvO) approach will result in the following:

Setosa against Versicolor
Setosa against Virginica
Versicolor against Virginica
Virginica against Versicolor.

Using the formula below, we can calculate the number of data divides.

Split data = (number of classes X (number of classes − 1))/2. Other aspects of this strategy are comparable to the one-versus-rest method. To implement this method, we can use the scikit-learn library, which has the OneVsOneClassifier method within the multi-class package.

1. **Let Us Import Libraries First**

```
from sklearn import datasets

from sklearn.svm import SVC

from sklearn.multiclass import OneVsOneClassifier

import matplotlib.pyplot as plt

from sklearn import metrics

from sklearn.metrics import classification_report, confusion_matrix, accuracy_score
```

2. Load the Iris Dataset and Apply the One-Versus-One SVM for Classification

```
iris =datasets.load_iris()

X=iris.data

y =iris.target

svc =SVC()

Train the model

OvO=OneVsOneClassifier(svc)

OvO.fit(X, y)

Predicting values from the model.

Result =OvO.predict(X)

Result
```

Output

```
array([0, 0, 0, 0, 0, 0, 0, 0, 0, 0, 0, 0, 0, 0, 0, 0, 0, 0, 0, 0, 0, 0,
       0, 0, 0, 0, 0, 0, 0, 0, 0, 0, 0, 0, 0, 0, 0, 0, 0, 0, 0, 0, 0, 0,
       0, 0, 0, 0, 0, 0, 1, 1, 1, 1, 1, 1, 1, 1, 1, 1, 1, 1, 1, 1, 1, 1,
       1, 1, 1, 1, 1, 1, 1, 1, 1, 1, 1, 2, 1, 1, 1, 1, 1, 2, 1, 1, 1, 1,
       1, 1, 1, 1, 1, 1, 1, 1, 1, 1, 1, 1, 2, 2, 2, 2, 2, 2, 1, 2, 2, 2,
       2, 2, 2, 2, 2, 2, 2, 2, 2, 2, 2, 2, 2, 2, 2, 2, 2, 2, 2, 2, 2, 2,
       2, 2, 2, 2, 2, 2, 1, 2, 2, 2, 2, 2, 2, 2, 2, 2, 2, 2])
```

3. Print the Confusion Matrix

```
mat2=confusion_matrix(y,Result)

print("Confusion Matrix:\n",t)

cmat2_display =metrics.ConfusionMatrixDisplay(confusion_matrix =mat, display_labels =[0,
1,2])

cmat2_display.plot()

plt.show ()
```

Output

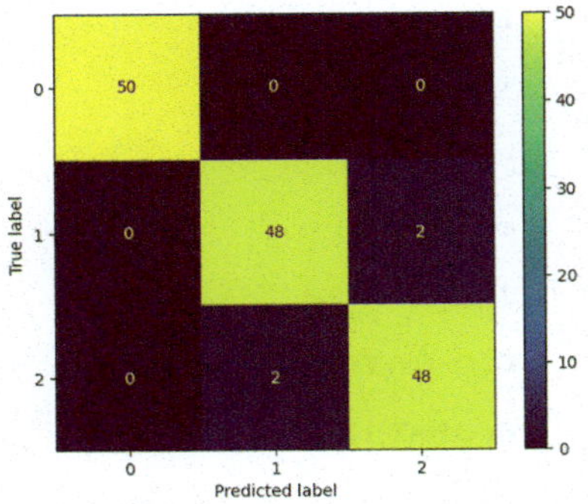

This example shows that SVM can predict more than two classes using the one-versus-rest and one-versus-one methods. We can also analyze from the confusion matrices that the one-versus-one method performs well compared to the one-versus-rest. However, we can easily swap it with other classifiers such as logistic regression, random forest classifier, or k-nearest neighbors. The steps for data preparation and evaluation remain the same.

3.5 Multi-class Classification with Imbalanced Dataset

Imbalanced data often describes a classification problem in which the classes are not equally represented. Consider a four-class classification issue for 100 fruits classified as grapes, bananas, apples, oranges. A total of 70 examples are designated as Class-1 (grapes), 10 instances as Class-2 (bananas), 10 instances as Class-3 (apples), and the remaining 10 instances as Class-4 (oranges). This dataset is unbalanced with an 8:1:1:1 ratio.

Imbalanced dataset categorization is a relatively recent research area within the larger context of machine learning studies, seeking to learn from skewed data distributions [8]. Most traditional machine learning methods perform poorly in these datasets because they prefer majority class samples, resulting in lower projected accuracy for the minority class. [9, 10]. There are mainly two ways to handle the class imbalance problem: data level and algorithmic level. Data-level approaches use resampling techniques to rebalance data distribution. These solutions tackle the problem by boosting or decreasing minority class observations. Many different sampling procedures have been developed, including random oversampling, random undersampling, SMOTE, direct oversampling, and other similar methods, each with merits and downsides. The algorithm-level approaches improve the machine learning algorithm's ability to handle unbalanced data. The adjustment may include calculating misclassification costs for each class, often known as cost-sensitive algorithms, and minimizing the cost error rather than optimizing the accuracy rate. The next sections illustrate various strategies for multi-class classification using an imbalanced dataset.

Data-Level Methods

1. **Resampling**: choosing or creating a certain number of instances from the majority or minority classes to rebalance them and reduce the impact of imbalance (undersampling, oversampling, and hybrid). Oversampling procedures can repeat existing samples or add new ones to the minority class. They expand the training set, requiring additional time. Undersampling approaches remove samples from majority classes until both minority and majority classes are evenly distributed. As a result, the training procedure is simple to implement and improves the problem of run time and storage [11]. For multi-class datasets, sampling strategies and procedures should be modified to account for specific class attributes and their interrelationships. Hybrid sampling approaches use many sampling techniques to address each technique's limitations.
2. **SMOTE**: This can lead to overgeneralization because the majority class is not considered. Because the created synthetic samples neglect the majority class, there may be class overlap, particularly in multi-class scenarios.

Algorithm-Level Methods

In the algorithm-level category, the existing learner is adjusted to eliminate its bias toward the majority classes. Here are some effective algorithmic methods:

1. **Cost-Sensitive Learning**: The most prevalent strategy is cost-sensitive learning, which requires the learner to correctly classify minority class samples by assigning a high penalty to misclassified minority class samples. There is no penalty for proper classification samples; however, minority samples have a higher misclassification cost than majority samples. The goal is to minimize the overall cost of the training dataset [12].
2. **Ensemble Methods**: Ensemble approaches integrate many models to increase performance and address class imbalances. Ensemble learning is one of the most commonly used classifiers that integrate data and algorithmic-level methods for dealing with imbalanced data problems [13].
3. **Boosting Algorithms**: Boosting algorithms like AdaBoost.MH, SAMME, LogitBoost, GradientBoost, XGBoost, LightGBM, CatBoost, SMOTEBoost, RUSBoost, MEBoost, AdaCost, AdaC1, AdaC2, and AdaC3 focus on difficult-to-classify instances, which can help handle class imbalance. Boosting approaches entail sequentially learning weak classifiers to reduce mistakes from

previous classifiers. Several boosting approaches have been presented, and for this work, I will use the CatBoost algorithm to make predictions.

4. **Class Weight Adjustment**: Many algorithms in scikit-learn allow for class weight adjustments, such as logistic regression, SVM, and decision trees.

3.6 Hands-On Experience with Multi-class Classification on Class Imbalance Dataset Using the Seattle Weather Dataset

Multi-class weather classification is a valuable technology with numerous applications, including video surveillance and intelligent transportation. However, the task is tough due to varying weather conditions and the unequal number of different weather conditions such as sunny, cloudy, rainy, and snowy in the dataset [14, 15]. To understand the multi-class classification on the imbalance dataset, we are using the Seattle weather dataset from Kaggle (https://www.kaggle.com/datasets/malihachaity/heuristic-model).

Seattle weather dataset has four attributes and five classes.

PRCP—Precipitation
TMAX—Maximum Temperature
TMIN—Minimum Temperature
RAIN—TRUE—If it rained that day, FALSE—If it did not rain that day
Class = drizzle, rain, sun, snow, and fog.

There are 1461 observations in the dataset, and the number of observations in each class is imbalanced as shown in Fig. 3.9.

- Class 1: drizzle: 53 examples
- Class 2: rain: 641 examples
- Class 3: sun: 17 examples
- Class 4: snow: 26 examples
- Class 5: fog: 101 examples.

It can be seen that the dataset contains two major classes (rain and sun) and three minority classes. However, all classes are equally important in this prediction problem.

Figure 3.9a, b represents the histogram of examples in each class in the Seattle weather multi-class classification dataset

Class	Count
drizzle	53
rain	641
sun	640
snow	26
fog	101

(a)

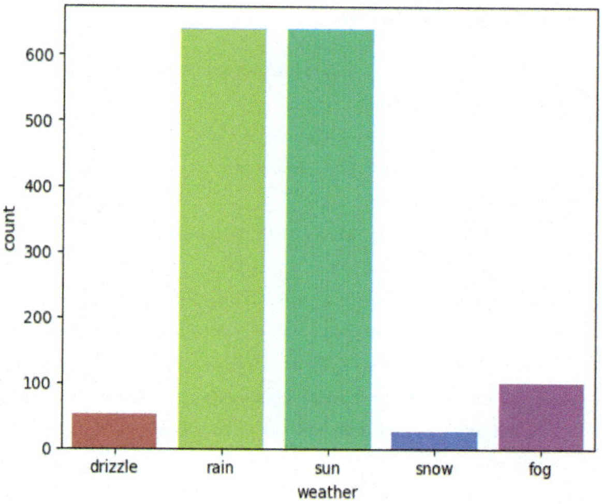

(b)

Fig. 3.9 Histogram of examples in each class in the Seattle weather multi-class classification dataset

Now, we will apply both data-level methods and algorithm-level methods for handling the class imbalance problem on this multi-class classification dataset. Oversampling is one of the most widely used techniques to deal with imbalance classes. Cost-sensitive learning is a subfield of machine learning that considers the costs of prediction errors (among other factors) when training a machine learning model. It is a subfield of unbalanced learning that studies classification on datasets with skewed class distributions. Apart from that, boosting algorithms which are a type of ensemble learning approach enhance the performance of individual base learners' in multi-class imbalanced learning.

(a) **Oversampling for Multi-class Classification with Imblearn**

Oversampling is the process of reproducing or synthesizing fresh examples of minority classes so that the number of examples in the minority class more closely approaches or matches that of the majority class. The ***imblearn*** library, which is part of the ***imbalanced-learn*** package, provides several methods for oversampling. Here is a general approach using the ***imblearn*** library.

Python Implementation

```
from sklearn.model_selection import train_test_split
# loading the dataset
df2 = pd.read_csv('/seattle.csv')
features = ['precipitation','temp_max','temp_min','wind']
label='weather'
X, y = df2[features].values, df2[label].values
# dividing X,y into train and test data
X_train, X_test, y_train, y_test = train_test_split(X, y, test_size=0.20, random_state = 0)
from imblearn.over_sampling import RandomOverSampler
from sklearn.ensemble import RandomForestClassifier
over=RandomOverSampler(sampling_strategy='auto',random_state=42)
X_resampled,y_resampled =over.fit_resample(X_train,y_train)
# apply Random Forest Classifier
classifier=RandomForestClassifier(random_state=42)
classifier.fit(X_resampled,y_resampled)
# Predicting the class
y_pred = classifier.predict(X_test)
accuracy = accuracy_score(y_test,y_pred)
precision = precision_score(y_test,y_pred,average="weighted")
Recall= recall_score(y_test,y_pred,average="weighted")
y_predprob = classifier.predict_proba(X_test)
print("Accuracy:",accuracy)
print("Precision:",precision)
print("Recall:",Recall)
```

Output

Accuracy: 0.7542662116040956
Precision: 0.7488941731673631
Recall: 0.7542662116040956.

(b) Penalize Algorithms (Cost-Sensitive Training)

Cost-sensitive training is a powerful strategy for dealing with imbalanced datasets, especially in multi-class classification tasks. This strategy involves altering the learning algorithm to account for the costs of misclassifying distinct classes. In general, ML algorithms seek to reduce error during training, and a variety of functions can be used to calculate a model's error or loss on training data. In cost-sensitive learning, a penalty is imposed for misclassifications, known as the cost. The goal of cost-sensitive learning is to reduce a model's misclassification cost based on input data. Thus, instead of optimizing accuracy, the algorithm aims to minimize the total misclassification cost [16, 17]. During training, we may use the option class_weight

= 'balanced' to penalize errors in the minority class by a factor proportional to how underrepresented it is. If we wish to use probability estimates in SVM algorithms, we should use the parameter probability = True.

Python Implementation

```
from sklearn.metrics import confusion_matrix
from sklearn.model_selection import train_test_split

# loading the iris dataset
features = ['precipitation','temp_max','temp_min','wind']
label = 'weather'
X, y = df2[features].values, df2[label].values

# dividing X, y into train and test data
X_train, X_test, y_train, y_test = train_test_split(X, y, test_size=0.20,
random_state = 0)

# training a linear SVM classifier
from sklearn.svm import SVC
clf = SVC(kernel='linear')
svc_model = SVC(class_weight='balanced', probability=True)

#train using train data
vm=svc_model.fit(X_train,y_train)

svm_predictions = svm.predict(X_test)

# model accuracy for X_test
accuracy = svm.score(X_test, y_test)
print(accuracy)
# creating a confusion matrix
cm = confusion_matrix(y_test, svm_predictions)
print (cm)
print("Precision:", precision_score(y_test, svm_predictions,
average="weighted"))
print('Recall:', recall_score(y_test, svm_predictions,
average="weighted"))
```

Output

Accuracy: 0.8122866894197952
Precision: 0.7286694478544159
Recall: 0.8122866894197952.

(c) **Ensemble Methods Using Random Forest Classifier**

While experimenting with a range of methods is a good rule of thumb for any machine learning task, it can be especially useful when dealing with imbalances [18, 19]. Decision trees usually perform effectively with unbalanced data. In modern machine learning, tree ensembles (random forests, gradient-boosted trees, etc.) almost always outperform single decision trees, so we will get right into them:

Python Implementation

```
from sklearn.ensemble import RandomForestClassifier
rfc =RandomForestClassifier()
# fit the predictor and target
rfc.fit(X_train,y_train)
# predict
rfc_predict=rfc.predict(X_test)# check performance
#print('ROCAUC score:',roc_auc_score(y_test,rfc_predict))
print('Accuracy score:',accuracy_score(y_test,rfc_predict))
print("Precision:",precision_score(y_test,rfc_predict,average="weighted"))
print('Recall:',recall_score(y_test,rfc_predict,average="weighted"))
```

Output

Accuracy score: 0.7781569965870307
Precision: 0.7483050498276739
Recall: 0.7781569965870307.

(d) **Boosting Algorithms**

Boosting algorithms are powerful techniques that can be very useful for dealing with multi-class unbalanced classification situations and boosting approaches work by sequentially training models, with each model attempting to rectify the faults of the previous one. Boosting techniques include AdaBoost, CatBoost, gradient boosting, and XGBoost. These algorithms can be modified to address class imbalances by introducing class weights or sampling methods into the boosting framework. Here, we will employ the CatBoost algorithm for making predictions.

Python Implementation

```
# load the libraries
from catboost import CatBoostClassifier
from sklearn.metrics import accuracy_score,precision_score,recall_score,
# Fit the model
CatModel=CatBoostClassifier()
CatModel.fit(X_train,y_train,verbose=256)
# Make predictions
catpreds=CatModel.predict(X_test)
y_pred_probs=CatModel.predict_proba(X_test)
print("Accuracy:",accuracy_score(y_test,catpreds))
print("Precision:",precision_score(y_test,catpreds,average="weighted"))
print('Recall:',recall_score(y_test,catpreds,average="weighted"))
```

Output

Accuracy: 0.7781569965870307
Precision: 0.7377894225906547
Recall: 0.7781569965870307.

It was found that cost-sensitive algorithms outperformed other algorithms in terms of accuracy and recall. Giving more weight to minority misclassifications and penalizing the model more for minority prediction errors leads to a model that pays more attention to this class. This forces the model to learn cases from the minority class, resulting in a model that is skilled at predicting that class.

3.7 Conclusions

The term "multi-class classification" refers to machine learning classification tasks involving more than two classes. In this chapter, we have presented a detailed overview of multi-class classification, bridging the gap from binary classification and addressing the specific issues and approaches applicable to multi-class cases. We presented theoretical insights and practical examples to demonstrate multiple strategies for effectively handling multi-class classification, even in the presence of class imbalances. This chapter provides readers with the knowledge needed to design robust multi-class classification models, allowing them to tackle complicated classification issues across multiple domains.

Bibliography

1. Kibria HB, Matin A (2022) The severity prediction of the binary and multi-class cardiovascular disease—a machine learning-based fusion approach. Comput Biol Chem 98:107672
2. Er MJ, Venkatesan R, Wang N (2016) An online universal classifier for binary, multi-class, and multi-label classification. In 2016 IEEE International Conference on systems, man, and cybernetics (SMC). IEEE, pp 003701–003706
3. Del Moral P, Nowaczyk S, Pashami S (2022) Why is multiclass classification hard? IEEE Access 10:80448–80462
4. Pawara P, Okafor E, Groefsema M, He S, Schomaker LR, Wiering MA (2020) One-vs-one classification for deep neural networks. Pattern Recogn 108:107528
5. Chaitra P, Kumar RS (2018) A review of multi-class classification algorithms. Int J Pure Appl Math 118(14):17–26
6. Xue Y, Hauskrecht M (2019) Active learning of multi-class classification models from ordered class sets. In: Proceedings of the AAAI conference on artificial intelligence, vol 33, No. 01, pp 5589–5596
7. Chamasemani FF, Singh YP (2011) Multi-class support vector machine (SVM) classifiers—an application in hypothyroid detection and classification. In: 2011 sixth international conference on bio-inspired computing: theories and applications. IEEE, pp 351–356
8. Tanha J, Abdi Y, Samadi N, Razzaghi N, Asadpour M (2020) Boosting methods for multi-class imbalanced data classification: an experimental review. J Big Data 7:1–47

9. Batista GE, Prati RC, Monard MC (2004) A study of the behavior of several methods for balancing machine learning training data. ACM SIGKDD Explorations Newsl 6(1):20–29
10. Lango M, Stefanowski J (2022) What makes multi-class imbalanced problems difficult? An experimental study. Expert Syst Appl 199:116962
11. Ling CX, Sheng VS (2008) Cost-sensitive learning and the class imbalance problem. Encyclopedia Mach Learn. 2011:231–235
12. Tahir MA, Asghar S, Manzoor A, Noor MA (2019) A classification model for class imbalance dataset using genetic programming. IEEE Access 8(7):71013–71377
13. Shim SO (2020) Multi-class classification based on relative distribution of class. In: 2020 2nd International conference on computer and information sciences (ICCIS)
14. Fenyi A, Asante M (2023) Multi-class weather classification using machine learning techniques. J Theor Appl Inform Technol 101(1)
15. Li Z, Li Y, Zhong J, Chen Y (2020) Multi-class weather classification based on multi-feature weighted fusion method. In: IOP conference series: earth and environmental science, vol 558, No. 4. IOP Publishing, p 042038
16. Hoens TR, Chawla NV (2013) Imbalanced datasets: from sampling to classifiers. Imbalanced learning: foundations, algorithms, and applications, pp 43–59
17. Mienye ID, Sun Y (2021) Performance analysis of cost-sensitive learning methods with application to imbalanced medical data. Inform Med Unlocked 25:100690
18. Parmar A, Katariya R, Patel V (2019) A review on random forest: an ensemble classifier. In: International conference on intelligent data communication technologies and internet of things (ICICI) 2018. Springer, Berlin, pp 758–763
19. Liu L, Wu X, Li S, Li Y, Tan S, Bai Y (2022) Solving the class imbalance problem using ensemble algorithm: application of screening for aortic dissection. BMC Med Inform Decis Mak 22(1):82

Chapter 4
Multi-objective and Multi-label Classification

4.1 Introduction

Conventional supervised learning algorithms operate in the context of a single label. In other words, each instance is linked to a single label out of a finite set of labels. On the other hand, in many real-world problems, every instance can be connected to multiple labels at the same time. One gene sequence may have more than one function in bioinformatics, while a single document may fall under several categories in text categorization. This is related to the challenge of assigning a set of labels to each occurrence; this is known as multi-label classification. The fields of data mining and machine learning have become more interested in multi-label classification in recent years.

A sophisticated machine learning technique known as "multi-objective classification" deals with several classification goals at once, frequently requiring trade-offs between opposing goals. Multi-objective classification takes into account multiple factors, including precision, recall, F1 score, computational complexity, and occasionally other domain-specific metrics, in contrast to standard single-objective classification, which concentrates on optimizing a single performance parameter like accuracy. This methodology proves to be especially advantageous in intricate real-world settings when attaining superior performance in one metric could result in less-than-ideal outcomes in another, therefore requiring a delicate balance. The Pareto front is a collection of ideal solutions that are obtained by navigating trade-offs between various objectives through the use of techniques like Pareto optimization, weighted sum methods, and evolutionary algorithms. With the help of these solutions, decision-makers can choose the model that best suits their requirements and budget. In industries like healthcare, banking, and autonomous systems, where misclassification can have serious repercussions and necessitate a sophisticated approach to model evaluation and selection, multi-objective classification is frequently used. The creation of diagnostic tools for identifying the existence of diseases like cancer is one instance of multi-objective classification in action in the healthcare sector. In this

S. Chakraborty and L. Dey, *Multi-objective, Multi-class and Multi-label Data Classification with Class Imbalance*, Springer Tracts in Nature-Inspired Computing,
https://doi.org/10.1007/978-981-97-9622-9_4

case, a model has to strike a compromise between two goals: maximizing the true negative rate (specificity) to reduce the number of healthy people who are mistakenly diagnosed with the disease and maximizing the true-positive rate (sensitivity) to guarantee that the majority of patients with the illness are identified correctly. Furthermore, reducing the false positive rate is critical to spare patients needless worry and medical interventions, and reducing the false negative rate is critical to guarantee prompt and precise care for individuals impacted. By using multi-objective classification techniques, a diagnostic model that offers a balanced performance across different parameters can be developed, providing medical professionals with a more comprehensive and dependable tool and improving patient outcomes [1, 2]. It has a list of benefits described below.

- **Balanced Optimization**: A more balanced model performance is achieved by using multi-objective classification, which enables the simultaneous optimization of several performance metrics, including precision, recall, and F1 score.
- **Trade-off Management**: It facilitates the management of trade-offs between competing goals by giving a better knowledge of the performance sacrifices that must be made in the model.
- **Pareto Efficiency**: With the help of several models that illustrate various trade-offs, this method produces a collection of Pareto-optimal solutions, providing decision-makers with a variety of possibilities to select from per certain requirements.
- **Increased Flexibility**: Multi-objective classification is appropriate for dynamic domains where priorities may fluctuate because it can adjust to changes in the relative importance of several objectives over time.
- **Comprehensive Assessment**: It encourages a comprehensive assessment of models, taking into account multiple performance factors as opposed to concentrating only on one metric, which results in more durable and dependable models.
- **Better Decision-Making**: This strategy facilitates more informed decision-making by helping to comprehend the connections and trade-offs between various goals.
- **Personalization**: It enables models to be tailored to the unique needs of various applications, increasing their applicability and efficacy.

Unlike traditional classification tasks, which limit an instance to a single label, multi-label classification uses a sophisticated machine learning technique that allows an instance to be simultaneously linked with numerous labels. This paradigm is especially helpful in situations where the data points simultaneously belong to different categories or show many, non-exclusive qualities. Multi-label classification has to deal with the difficulty of predicting a subset of all potential labels, in contrast to single-label classification, which aims to predict one class from a set of mutually incompatible classes. Multi-label classification techniques encompass issue transformation approaches like binary relevance and classifier chains, as well as algorithm adaptation techniques that involve modifying pre-existing algorithms to accommodate multiple labels. In addition to being different from single-label classification, multi-label classification evaluation metrics use metrics including F1 score over multiple labels, precision, recall, and Hamming loss. This approach has wide applications in fields, such as image recognition (where an image may contain many objects), text categorization (where a document may refer to multiple subjects), and bioinformatics (where genes may perform multiple roles). Multi-label categorization offers a more thorough and nuanced comprehension and prediction capabilities in complicated areas by addressing the diverse character of data. Multi-label classification in action can be observed in email categorization systems utilized by providers, such as Gmail. Emails can concurrently fall under several categories, including “Work,” “Personal,” “Promotions,” “Social,” and “Updates.“ An email regarding a work social event, for example, may be filed under both “Work” and “Social.“ Each email is analyzed by the multi-label classification system, which then assigns several pertinent labels to make sure the user’s inbox is organized correctly. Email organization and accessibility are enhanced by techniques like binary relevance and classifier chains, which manage the difficulty of assigning many labels. By guaranteeing that emails are correctly classified under all relevant categories, our method improves user experience and makes email management and retrieval more efficient [3, 4]. There are several advantages to using multi-label classification, especially in intricate and varied real-world applications:

- **Handling Complex Data**: To capture the complexity of real-world data, Multi-label Classification works best in situations where examples belong to several categories at once.
- **Enhanced Information Extraction**: By anticipating numerous labels for each instance, it extracts more information, resulting in a more comprehensive comprehension of the data.
- **Efficiency in Data Processing**: It expedites data processing and boosts organizational efficiency by simultaneously classifying data into several pertinent labels.
- **Enhanced User Experience**: By precisely classifying objects under all relevant categories, it improves user experience in applications, such as email classification and recommendation systems.
- **Versatility**: Multiple labels per instance are frequent in many disciplines where Multi-label Classification can be used, including text categorization, image recognition, and bioinformatics.
- **Improved Resource Utilization**: Compared to running several single-label classifiers, it optimizes the usage of computing resources by addressing many labels at once.
- **Enhanced prediction Power**: It captures the multidimensional structure of the data by utilizing the interdependencies between labels to increase overall prediction performance.

In this chapter, a detailed discussion of multi-objective and multi-label classification techniques with some interesting case studies of different application areas is presented.

4.2 Difference Between Multi-class and Multi-label Classification

The issue definitions, applications, and assessment metrics of multi-class and multi-label classification are where the main distinctions between them are found.

A. **Defnition**

Multi-class classification: In this kind of classification, each instance is assigned to a single class from a group of mutually exclusive classes, classifying animal species, for instance, so that each animal is only a dog, cat, or bird at a time.

Multi-label classification: In this case, an instance may concurrently belong to several classes. An instance can have more than one label assigned to it, and the classes are not exclusive. An email, for example, can have the tags "Work" and "Urgent."

B. **Formulation of the Problem**

Multi-class classification: Predicting a single label from a group of different classes is the aim. In the last layer of a neural network, the problem is often presented with a single output node that represents the probability of each class.

Multi-label classification: Predicting a collection of labels for every case is the aim. In the last layer of a neural network, the problem is presented with numerous output nodes, where each node reflects the likelihood of a label being present.

C. **Areas of Applications**

Multi-class classification: It is frequently employed in situations where each data point belongs to one category, such as in the classification of species in biology, sentiment analysis (positive, negative, or neutral), and handwritten digit identification (0–9).

Multi-label classification: Applicable in situations where data points can be assigned to more than one category. Examples include text classification (sports, politics, and technology), image tagging (a beach, a sunset, and people can be found in an image), and music genre classification (a song can fall under the categories of rock, blues, and jazz).

D. **Metrics for Evaluation**

Multi-class classification: When comparing a single predicted label to the true label, evaluation metrics like as accuracy, precision, recall, F1 score, and confusion matrix are generally computed.

Multi-label classification: In multi-label classification, evaluation metrics are computed by comparing the set of predicted labels with the set of true labels. These metrics are more complicated and include recall, F1 score, precision, recall, Hamming loss, and Jaccard index.

Fig. 4.1 Classification of single-label images

E. **Computational Complexity**

Multi-class classification: As each instance belongs to a single class, it is typically less complicated. Training might be easier and the models frequently simpler.

Multi-label classification: More difficult since each instance requires the prediction of multiple labels. This necessitates managing label correlations and frequently calls for increasingly complex algorithms and assessment methods.

The majority of us mix up image classification with multi-label and multi-class methods. In this section, the differences are going to be cleared through some examples. Assume we are presented with animal images (Fig. 4.1) that need to be categorized into appropriate groups. For the sake of clarity, let us say that an image can be classified into a total of 4 categories: cat, dog, rabbit, and parrot. There are now two possible outcomes:

A. Since each image only includes one object (from the four categories listed above), it can only be assigned to one of the four groups.
B. The image may include more than one object (from the four categories mentioned above), in which case it will fall under multiple categories.

The images below include just one item in them. Those with severe eyes will have identified four distinct categories of objects (animals) within this collection. There is only one type of image that fits this category: cat, dog, parrot, or rabbit. An image cannot be included in more than one category.

1. When the images can be categorized into more than two groups.
2. An image can only be included in one category.

An image classification problem is said to be as multi-class if each of the aforementioned requirements are met. Let us now examine the second case; have a look at Fig. 4.2.

Fig. 4.2 Classification of multi-label images

There is a dog and a cat in the first image (top left) of Fig. 4.2. A dog, a cat, and a parrot may be seen in the second shot (top right). There is a rabbit and a parrot in the third image (bottom left). In the final image (bottom right), a dog and a parrot are shown. The labels for the provided images are all these. This is a multi-label image classification problem since each image in the set belongs to multiple classes.

4.3 Multi-objective SVM for Multi-class Classification

Support vector machine (SVM) is a supervised machine learning technique that is mainly used for classification tasks. SVM operates by locating, in a high-dimensional space, the hyperplane that optimally divides data points into discrete classes. This is accomplished by increasing the margin, or the separation between each class's nearest data point and the hyperplane, which strengthens the model's capacity for generalization. By using kernel functions, such as polynomial or radial basis function (RBF) kernels, which implicitly transfer input characteristics onto higher-dimensional spaces where a hyperplane can be utilized for separation, the approach can handle nonlinearly separable data. Strong variable classifiers (SVMs) are renowned for their resilience, capacity to manage high-dimensional data, and efficiency in scenarios when the quantity of dimensions surpasses the quantity of samples. The separation hyperplane is shown by the solid red line in Fig. 4.3, and the support vectors are represented by the three points—two blue and one green—that are distributed over the lines.

Multiclass classification is not natively supported by SVM. It facilitates the division of data points into two classes and binary classification. After decomposing the multi-classification problem into several binary classification problems, the same method is applied to multiclass classification. Optimizing two or more

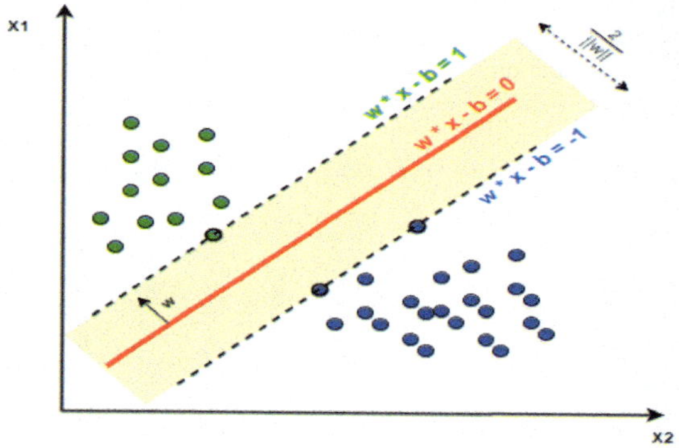

Fig. 4.3 SVM using hyperplane for basic classification

competing objectives at the same time is known as multi-objective optimization. These goals could involve maximizing margin, lowering model complexity, and minimizing classification error in the context of SVMs. By attempting to strike a balance between these conflicting objectives, multi-objective optimization frequently produces a group of Pareto front optimal solutions [5, 6].

Multiple Classification: Due to their inherent ability to discriminate between two classes, traditional support vector machines (SVMs) are binary classifiers. Common methods for extending SVMs to multi-class classification are as follows:

One-versus-One (OvO): For each pair of classes, create a binary classifier. Voting schemes are frequently used to determine the final classification. For k classes, this results in $k * (k - 1)/2$ classifiers. In Fig. 4.4, the sole goal of the red-blue line is to maximize the distance between the blue and red spots. Green points are unrelated to it.

One-versus-All (OvA): Builds a binary classifier that compares each class to every other class. This yields k classifiers for k classes. Usually, the classifier with the highest confidence score determines the final categorization. In Fig. 4.5, the green

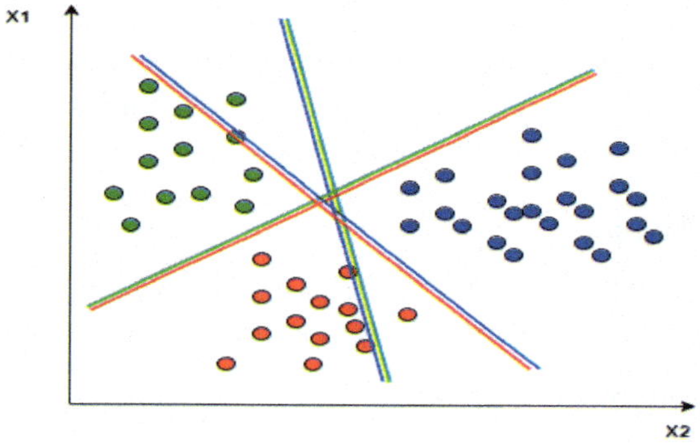

Fig. 4.4 One-versus-one SVM approach for three classes separation

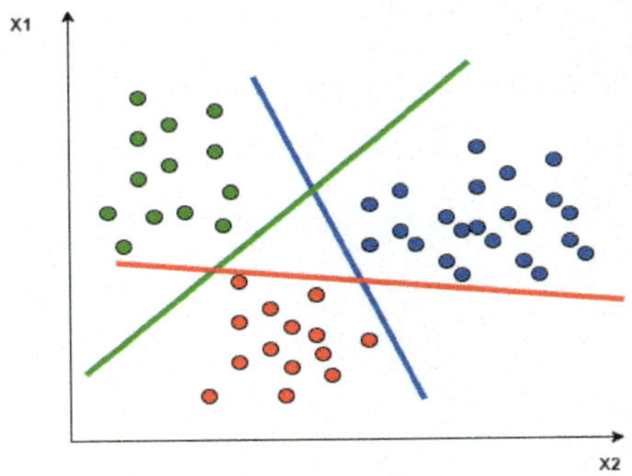

Fig. 4.5 One-versus-All SVM approach for three classes separation

line seeks to simultaneously maximize the distance between green points and every other point.

Integrating multi-objective optimization with multi-class SVM involves several steps. Establish several goals for optimization. Typical goals consist of:

1. **Classification Error:** Reduce the number of cases that are incorrectly classified. Maximize the margin between classes to enhance generalization. Reduce the complexity of the model as much as possible; this may entail regularization terms.
2. **Integrate Multi-class SVM with Multi-objective Optimization:** Utilize methods, such as One-vs-One or One-vs-All when classifying many classes. To determine the optimal trade-offs between the specified objectives, use multi-objective optimization.
3. **Methods of Optimization:**

   - **Pareto optimization**: Produce a collection of Pareto-optimal solutions using Pareto optimization. Every option reflects a unique trade-off between the goals.
   - **Weighted Sum Method:** Assign weights to each aim to combine them into a single scalar objective. Maximize the combined goal.
   - **Evolutionary Algorithms:** To find Pareto-optimal solutions, use algorithms, such as NSGA-II (Non-dominated Sorting Genetic Algorithm II).
   - **Evaluation and Selection:** Choose the best model by considering the intended trade-offs and evaluating the solutions using the relevant metrics (e.g., accuracy, precision, recall).

Example

Let us look at an illustration of a multi-objective SVM with a One-versus-All strategy utilizing the weighted sum approach for multi-class classification:

1. **Specify the Goals:**

Goal 1: Reduce the amount of categorization error (E).

Goal 2: To maximize the margin (M).

Goal 3: Reduce the complexity of the model (C).

2 **Weighted Sum Objective**: Consolidate goals into a single-objective function using a weighted sum objective:

$$F = w_1 \cdot E + w_2 \cdot \frac{1}{M} + w_3 \cdot C$$

where, w_1, w_2, and w_3 are the weights associated with each objective.

3. **Optimization**: Apply an algorithm for optimization (e.g., gradient descent, evolutionary algorithm) to optimize F.
4. **Train SVMs**: Using the optimized objective function, train binary SVMs for every class.
5. **Evaluate and Choose**: Based on the combined objective function and intended trade-offs, evaluate the models on a validation set and choose the optimal model.

Multi-class support vector machines (SVMs) can be trained with multi-objective optimization to attain balanced performance that fits your application's unique requirements and limitations [5].

4.4 Multi-objective and Multi-label Feature Selection

One of the primary preprocessing steps in machine learning is feature selection, which can remove *k* high-efficiency features from original feature sets containing n features ($k < n$). Algorithm cost and running time can be effectively reduced by choosing features carefully. Generally speaking, there are three types of feature selection methods: wrapper, filter, and embedding approaches. The filter approach begins by identifying the optimal feature subset within the dataset, after which a classifier is trained. Because of its pertinence, the wrapper technique frequently obtains more applicable classifiers than the filter method, which can be thought of as the benchmark for feature selection. However, the wrapper method takes longer and costs more money because it requires more classifier training. The embedded approach optimizes both feature selection and classifier using a single algorithm.

4.4.1 Multi-objective Feature Selection

A powerful machine learning technique called multi-objective feature selection (MOFS) simultaneously optimizes several competing objectives to choose the most pertinent subset of characteristics from a dataset. Multi-objective feature selection balances multiple objectives, such as improving model performance, minimizing the number of features, reducing redundancy among features, and maximizing relevance to the target variable, in contrast to traditional feature selection methods that concentrate on a single criterion, such as maximizing accuracy. To evolve feature subsets, this technique entails setting various objectives, such as maximizing prediction accuracy and minimizing feature count, and using optimization algorithms, such as particle swarm optimization (PSO), genetic algorithms (GA), or NSGA-II. Rather than giving decision-makers a single ideal subset, the method finds a Pareto front, which stands for feature sets that provide the best trade-offs between the objectives. This methodology is especially useful in complex applications where great accuracy must be achieved while preserving simplicity and interpretability, like medical diagnostics. Multi-objective optimization helps build reliable, effective, and understandable models by addressing several facets of feature selection. This eventually improves the functionality and performance of machine learning systems [7]. The objectives in multi-objective feature selection are given below,

- **Optimizing Model Performance**: This entails choosing characteristics that improve the model's recall, precision, accuracy, or other pertinent performance measures.
- **Minimizing Feature Set Size**: Cutting down on features makes the model easier to understand, requires less computing power, and is more readable.
- **Reducing Redundancy**: Making sure that certain aspects do not have a strong correlation or are redundant with one another to preserve important and distinct information.
- **Maximizing Relevance**: Choosing attributes with the greatest significance or relevance to the goal variable is known as "maximizing relevance."

The goal of MOFS for a classification assignment is to identify a set of trade-off solutions between a learner's classification error rate and the number of selected features that are minimized. Given a dataset with D characteristics and L instances, the following is how a MOFS problem may be expressed:

$$\begin{aligned} \text{minimize} \quad & \mathbf{F}(\mathbf{X}) = (f_{\text{ratio}}(\mathbf{X}), f_{\text{error}}(\mathbf{X}))^T, \\ \text{s.t.} \quad & \mathbf{x} = (x_1, \ldots, x_D)^T, \end{aligned}$$

where x is a solution to binary encoding that encodes a subset of features; for $i = [1, \ldots, D]$, $x_i = 0$ and $x_i = 1$ indicate selection and discarding of the ith feature, respectively. There are two goals: $f_{\text{error}}(x)$ and $f_{\text{ratio}}(x)$. When comparing solutions, the idea of Pareto dominance is crucial. Only when x is superior to y in at least one objective and equal to y in all objectives does x outperform y in the other solution. A solution is deemed Pareto-optimal if it dominates x and no other solution does. The Pareto set (PS) is the collection of all Pareto-optimal solutions.

The ratio of the number of selected features to the total number of features can be used to illustrate the first objective fratio, which is to minimize the number of selected features.

$$f_{\text{ratio}}(\mathbf{X}) = \frac{\sum_{i=1}^{D} x_i}{D}$$

The second objective ferror, which is to minimize the classification error rate, is more challenging than the first. Generally, a particular classifier assesses the classification error associated with a group of features. The following is a formulation of f_{error},

$$f_{\text{error}}(\mathbf{X}) = 1 - \frac{1}{c}\sum_{i=1}^{c} \frac{TP_i(\mathbf{x})}{|S_i|}$$

where $|S_i|$ is the number of instances of class i, c is the number of classes in a dataset, and $TP_i(x)$ is the number of successfully detected instances in class i when utilizing x for classification.

4.4.2 Multi-label Feature Selection

An advanced machine learning technique called multi-label feature selection is designed for situations when an instance can be assigned to more than one label at the same time. Its goal is to identify the most informative characteristics that apply to all labels. Multi-label feature selection necessitates taking into account the correlations and interdependencies across several labels, in contrast to conventional feature selection techniques meant for single-label situations. Through the use of techniques like mutual information, chi-square, or specialized multi-label algorithms like multi-label relief or multi-label feature selection via dependence maximization, features are assessed for significance depending on how well they predict multiple labels. Finding a subset of characteristics that not only increases the model's predictive accuracy across all labels but also lowers computational cost and improves interpretability of the model is the aim. Through the emphasis on characteristics that have significance across many labels, this methodology guarantees a more thorough comprehension of the data and produces multi-label classifiers that are more proficient and successful. Applications like text categorization, where documents may fall under numerous subjects, and image annotation, where images include multiple objects, benefit greatly from multi-label feature selection, which eventually leads to more relevant and accurate predictions [8].

4.4.2.1 Methods for Multi-label Feature Selection

1. **Filter Techniques**

Mutual Information: Determines the mutual information shared by all labels and each feature together. Higher mutual information features concerning the set of labels are deemed more significant.

Chi-square: By averaging the chi-square values for each label, this test extends its application to multi-label data.

2. Wrappers Techniques

Multi-label Recursive Feature Elimination (ML-RFE): This methodology extends the recursive feature elimination method by assessing features according to how they affect the multi-label classifier's performance.

Genetic Algorithms: By comparing several label-based fitness functions, evolutionary algorithms are used to find the best subsets of attributes.

3. **Embedded Approaches**

Multi-label Decision Trees: Uses tree-based algorithms that automatically choose pertinent features based on multiple label splits to incorporate feature selection into the model training process.

Regularization Techniques: Integrates label correlations directly into the optimization target, allowing regularization techniques like Lasso (L1) or Elastic Net to handle multiple labels.

Example Workflow

(1) **Data preprocessing**: To make sure that the features are on the same scale, normalize or standardize the dataset.
(2) **First Feature Selection**: To narrow down the feature space, use a filtering technique, such as mutual information. Choose the most pertinent features by comparing their mutual information scores with the labels.
(3) **Application of Wrapper approach**: Based on the performance of a multi-label classifier, employ a multi-label wrapper approach, such as ML-RFE, to iteratively remove the least significant features to further refine the feature set.
(4) **Model Training and Evaluation**: Using the chosen features, train a multi-label classifier (such as multi-label k-nearest neighbors or random forest) and assess its performance using metrics like precision, recall, and hamming loss.

Enhancing the effectiveness and efficiency of multi-label classification jobs requires multi-label feature selection. Feature selection techniques can find the most useful features by taking into account the relationships between labels, which results in more precise and understandable models. Mutual information, ML-RFE, and regularization techniques are useful tools in the machine learning toolbox because they provide reliable ways to address the challenges posed by multi-label data.

4.5 Multi-label Classification on Image

The process of multi-label classification on images includes assigning each image many labels that simultaneously indicate the existence of different items or features. By identifying numerous aspects inside a single image, multi-label classification tackles more complex circumstances than single-label classification, which assigns a single label to each image. In a binary relevance framework, where each label is handled as an independent binary classification problem, basic machine learning algorithms for this task include using techniques like logistic regression, k-nearest neighbors, or decision trees. Random forest is extensively used for multi-label image data classification. Although more sophisticated models like convolutional neural networks (CNNs) are frequently chosen for their superior performance on image data [9].

4.5.1 Techniques and Models

1. **Techniques for Transforming Problems**
   A. **Binary Relevance (BR):** Each label is handled as a distinct binary classification problem. Although this method is straightforward, label dependencies are not captured.
   B. **Classifier Chains (CC):** Using a sequential prediction process, each classifier in the chain takes into account the predictions of its predecessors as new input characteristics, thereby capturing label dependencies and extending Bayesian regression.
   C. **Label Powerset (LP):** Converts the problem into a multi-class classification problem by treating every distinct combination of labels as a single class. The exponential rise of label combinations makes this strategy unfeasible for huge numbers of labels.
2. **Algorithm Adaptation Methods**
   A. **Multi-label k-Nearest Neighbors (ML-kNN):** This technique counts the labels in the neighborhood and assigns labels based on a threshold, extending the kNN method to multi-label classification.
   B. **Multi-label Decision Trees:** Decision trees are handled by several labels. These trees are used to manage multi-label data by usually adjusting the splitting criterion.
   C. **Deep Learning Models:** By employing a binary cross-entropy loss function for each label and a sigmoid activation function for the output layer, convolutional neural networks (CNNs) can be modified for multi-label classification [10].

4.5.2 *Procedure of Multi-label Image Classification*

(1) **Obtain data in a structured format**: This can be used for multi-class and binary image categorization. All of the images you wish to use to train your model should be in a folder. We now need the true labels of the images to train this model. As a result, you want to also have a.csv file with the titles of all the training images together with their accurate labels.

(2) **Preprocessing**: Load each image from the collected dataset, then preprocess it under the specifications of our project. A validation set of images is constructed in order to assess the performance of our classification model using unobserved data, or test data. The validation set of images is used to verify our model after it has been trained on the training set.

(3) **Define model architecture and Training**: Defining the model's architecture is the next stage. A variety of models, such as VGGNet, ResNet, MobileNet, AlexNet, and vanilla CNNs, are appropriate for this task. Dividing an image into layers involves determining the number of hidden layers, neurons in each layer, the activation function, and other relevant factors.

We must train our model on the training set after it has been created and defined! To train the previously built model, we feed it training images along with the labels and classes that correspond to them. Here, we also send the validation images, which help to confirm how well the model works with untested data. When building the model, we use the RMSprop procedure to determine the root mean square values. At this point, we pass the validation images that verify the model's performance on untested data.

(4) **Prediction of multi-label images**: Predictions on test images are obtained using the trained model. Make a generator prediction with a probability greater than 0.7. That is, an image is classified as multi-label if the projected value is at least 70% closer to the original values.

Multi-label image classification is a difficult but necessary task for applications where images can be classified into more than one category. Practitioners can create strong models that can handle the complexity of multi-label data by utilizing strategies including issue transformation, algorithm adaptation, and deep learning. Effective and dependable multi-label image categorization systems are developed by careful study of label dependencies and suitable assessment criteria.

4.6 Multi-objective Optimization in Image Segmentation

To produce high-quality segmentation results, multi-objective optimization in image segmentation entails simultaneously optimizing numerous competing criteria. This method acknowledges that image segmentation tasks frequently necessitate striking a balance between multiple aspects, including computing efficiency, smoothness, and accuracy. For example, it is critical to precisely identify anatomical features in medical imaging while preserving smooth borders and reducing processing time. To identify a collection of optimal solutions that trade off multiple objectives, multi-objective optimization techniques are used, including Pareto-based approaches, genetic algorithms, and particle swarm optimization. These techniques let decision-makers select the best segmentation depending on the particular needs of the application by producing a Pareto front of potential options. This methodology improves the adaptability and efficiency of image segmentation in a variety of domains, such as computer vision, remote sensing, and medical imaging, by offering strong solutions that satisfy several requirements at once. To provide segmentation results that are more accurate and relevant, multi-objective optimization in image segmentation entails simultaneously optimizing numerous criteria. This strategy is critical for applications where high-quality segmentation is required, like object recognition, autonomous driving, and medical imaging [1, 11].

4.6.1 Main Objectives in Image Segmentation

- **Accuracy**

Reducing the amount of incorrectly identified pixels while guaranteeing that the segmented areas accurately depict the real objects in the image.

- **Smoothness**

Encouraging continuous, smooth borders to prevent noisy or jagged edges is crucial, especially for applications requiring clean anatomical boundaries, like medical imaging.

- **Compactness**

To prevent excessively fragmented or false sections that do not relate to meaningful objects, simpler, more compact shapes should be used for segmented regions.

- **Maintaining Topology**

Preserving an object's topological characteristics, which are vital for applications like biological cell segmentation and include things like the number of holes and related components.

- **Efficiency of Computation**

Maximizing the segmentation procedure to achieve computational viability, particularly for real-time uses like driverless vehicles.

Example Workflow

1. **Problem Definition**: Define the multiple objectives for segmentation, such as maximizing boundary accuracy and minimizing computational time.

2. **Data Preprocessing**: Prepare the image data, including noise reduction and contrast enhancement.
3. **Segmentation Algorithm**: Implement an algorithm (e.g., graph cuts, region growing, or deep learning) capable of incorporating multiple objectives.
4. **Optimization**: Apply a multi-objective optimization technique (e.g., NSGA-II) to find a set of optimal segmentation solutions.
5. **Evaluation**: Assess segmentation quality using metrics like the Dice coefficient for accuracy and boundary smoothness metrics.
6. **Selection**: Involve decision-makers in selecting the best segmentation result from the Pareto front based on specific application requirements.

Case Study

Multi-objective optimization can greatly improve the segmentation of intricate anatomical features in medical imaging. For instance, the following goals might be included in the segmentation of brain tumors from MRI scans:

- **Accuracy:** maximizing the overlap between the ground truth (as annotated by medical professionals) and the segmented tumor region.
- **Smoothness:** To make surgical planning easier, make sure the tumor margins are smooth.
- **Topology Preservation:** Preserving the continuity of brain structures and other anatomical accuracy.

The segmentation algorithm may produce high-quality, clinically valuable segmentation results that satisfy several requirements necessary for diagnosis and treatment planning by implementing multi-objective optimization [2].

4.6.2 Multi-objective Optimization Techniques

Finding solutions to several competing objectives is the goal of multi-objective optimization techniques, which are frequently applied to real-world issues requiring trade-offs. Using these strategies, decision-makers can come up with well-rounded solutions that sufficiently address all goals, producing more successful and long-lasting results in a variety of industries, including engineering, finance, and health care [12]. Here, several well-known multi-objective optimization methods are covered.

A. **Pareto Optimization**

In image segmentation, Pareto optimization involves determining a collection of solutions that optimally balance several competing objectives, including accuracy, smoothness, and processing efficiency. When dealing with complicated image segmentation jobs, where a single-objective optimization might not be enough to produce satisfactory results, this method is quite helpful. This method is especially helpful in domains where precise analysis and decision-making depend on high-quality segmentation, such as medical imaging, autonomous vehicles, and object recognition. Practitioners can attain superior segmentation performance tailored to the particular requirements of their applications by utilizing Pareto optimization approaches. Several algorithms and techniques are used to perform Pareto optimization in image segmentation [13–15]. Here are some of the most prominent ones.

A.1 NSGA-II (Non-dominated Sorting Genetic Algorithm II)

For image segmentation, Non-dominated Sorting Genetic Algorithm II (NSGA-II) is a potent multi-objective optimization method that skillfully strikes a balance between several segmentation requirements, including accuracy, smoothness, and computational efficiency. NSGA-II uses a rapid non-dominated sorting technique to categorize solutions into distinct Pareto fronts while also evolving a population of viable segmentation solutions through processes of crossover, mutation, and selection. The algorithm ensures a well-spread Pareto front by maintaining diversity among solutions through the application of a crowding distance technique. Due to its ability to produce a wide range of excellent segmentation results, NSGA-II is especially useful for difficult applications like medical image segmentation, where accurate anatomical structure delineation is crucial. NSGA-II offers a strong and adaptable framework for attaining excellent image segmentation performance catered to particular application requirements by concurrently maximizing several objectives [16].

A.2 SPEA2 (Strength Pareto Evolutionary Algorithm 2)

The sophisticated multi-objective optimization method known as Strength Pareto Evolutionary Algorithm 2 (SPEA2) is used in image segmentation to balance many segmentation goals, including accuracy, boundary smoothness, and processing economy. To maintain a well-distributed Pareto front, SPEA2 integrates both dominance and density information, assigning a strength value to each solution based on the number of solutions it dominates. Using a separate archive to hold non-dominated solutions and maintain diversity, the algorithm uses selection, crossover, and mutation processes to create a population of segmentation solutions. This approach produces a wide range of optimal solutions that address various trade-offs between conflicting criteria, making it very useful in medical imaging and other applications needing accurate and dependable segmentation. SPEA2 is a reliable and adaptable method for obtaining excellent image segmentation results that cater to particular requirements because of its capacity to tackle complicated multi-objective challenges.

B. **Weighted Sum Approach**

The weighted sum approach is a multi-objective optimization method for the segmentation of images that combines several criteria into a single-objective function by giving each criterion a certain weight. This technique reduces the complexity of the optimization process by treating it as a single-objective issue. The weighted total of the different objectives—accuracy, smoothness, and computational efficiency—makes up the overall objective function. Practitioners can establish an ideal balance that meets the unique requirements of the segmentation task by varying the weights, which in turn controls the trade-offs between various objectives. For instance, in medical imaging, more weight can be given to smoothness and precision to guarantee accurate and clinically significant segmentation of anatomical structures, while computing efficiency may be given priority in real-time applications. The weighted sum approach is very helpful because it is straightforward to apply, enabling flexible adaption to a range of image segmentation difficulties.

C. **Epsilon-Constraint Method**

The Epsilon-Constraint Method is a multi-objective optimization technique used in image segmentation. It takes a multi-objective problem and breaks it down into a series of single-objective problems by treating the other objectives as constraints with upper bounds, or "epsilon values," and optimizing only one primary objective. With the help of this technique, one may precisely regulate the trade-offs between various segmentation criteria, for example, optimizing segmentation accuracy while limiting computing time and border smoothness to reasonable levels. The approach may produce a wide range of high-quality segmentation solutions that approximate the Pareto front by iteratively modifying the epsilon values. This allows practitioners to choose the best option depending on their application needs. This method works especially well for complicated segmentation jobs, like medical imaging, where it is important to strike a balance between computing efficiency and precise anatomical structure delineation.

D. **Multi-objective Particle Swarm Optimization (MOPSO)**

An enhanced optimization method for image segmentation called multi-objective particle swarm optimization (MOPSO) expands on the concept of traditional particle swarm optimization by allowing it to handle multiple competing objectives at once. Within a repository of non-dominated solutions, a swarm of particles in MOPSO searches the solution space, guided by global and personal best positions. The trade-offs between goals like segmentation accuracy, border smoothness, and processing economy affect how each particle moves [17]. MOPSO can produce segmentation results of excellent quality by efficiently balancing these goals by keeping a wide range of Pareto-optimal solutions. This method is especially helpful for applications where accurate and effective segmentation is essential, such as remote sensing and medical imaging. Because of its adaptive nature, MOPSO may dynamically modify and improve its results, which makes it a reliable and adaptable method for dealing with challenging image segmentation problems.

E. **Genetic Algorithms (GAs)**

Genetic algorithms (GAs) optimize segmentation by generating a population of possible solutions. They do this by using evolutionary principles influenced by natural selection. Every member of the population is an encoded segmentation solution represented by a chromosome. GAs search the solution space iteratively to maximize various goals including accuracy, smoothness, and computational efficiency through processes of crossover, mutation, and selection. A variety of Pareto-optimal solutions are frequently maintained by non-dominated sorting and crowding distance methods, which guarantee a balance between various segmentation criteria. By offering a strong and flexible framework that progresses toward superior solutions, this method excels at managing the intricacy of image segmentation problems, such as distinguishing minute anatomical components in medical imaging or segmenting items in chaotic situations [18–20].

F. **Goal Programming**

Goal programming is a multi-objective optimization technique used in image segmentation that converts numerous segmentation objectives into distinct goals to minimize deviations from these intended targets. Every target, including computing efficiency, boundary smoothness, and segmentation accuracy, is given a goal that may also indicate a priority level, indicating how important it is. Goal programming makes it feasible to systematically balance conflicting objectives by framing the problem in this manner. This ensures that the most important goals are satisfied as closely as possible while also taking into account secondary purposes. This method is very helpful for applications, such as medical imaging, where accurate anatomical structure segmentation is crucial, but computational viability must also be maintained. This method is very helpful for applications, such as medical imaging, where accurate anatomical structure segmentation is crucial, but computational viability must also be maintained.

G. **Multi-objective Differential Evolution (MODE)**

An evolutionary technique called Multi-objective Differential Evolution (MODE) for image segmentation expands on the concept of traditional differential evolution by handling many competing objectives at once. A population of potential solutions is subjected to repeated processes of crossover, mutation, and selection in MODE to optimize multiple segmentation parameters, including computational efficiency, accuracy, and border smoothness. By employing Pareto dominance and diversity preservation techniques, the algorithm preserves a diverse variety of Pareto-optimal solutions, guaranteeing a well-distributed group of excellent segmentation outcomes. Due to its strong exploration and exploitation capabilities, MODE is especially useful for difficult image segmentation tasks, such as correctly detecting objects in diverse sceneries or precisely outlining anatomical structures in medical imaging. By striking a balance between several goals, MODE offers a strong and adaptable method for obtaining superior segmentation catered to certain application requirements.

H. **Interactive Methods**

To improve and refine results, interactive approaches for image segmentation actively incorporate user input or feedback during the segmentation process. By combining human knowledge and artificial algorithms, these techniques enable users to direct the segmentation process by making edits or remarks on preliminary results. Users can iteratively influence the segmentation process using techniques like progressive refinement and interactive evolutionary algorithms, ensuring that the final product satisfies individual needs and preferences. This method is especially useful for applications, such as medical imaging, where precise delineation of intricate anatomical structures requires professional expertise. Interactive approaches improve image segmentation accuracy, dependability, and customization by incorporating user input. This makes them appropriate for tasks that need high precision and flexibility in various settings.

4.7 Hands-On Experience on Real-Life Datasets

In this section, we present a straightforward approach to utilizing Python frameworks and toolsets for performing multi-label classification and multi-objective segmentation on images. Here, we are mainly focusing on the convolutional deep neural networks for this classification and segmentation task.

4.7.1 Multi-label Image Classification

First, set up the problem statement. We aim to identify a film's genre based solely on its poster artwork. What makes this a multi-label image classification task, and can you predict why? A film can indeed be in more than one genre. It is not limited to falling into an action, horror, or comedy subcategory. The film may blend elements of two or more different genres. Multi-label image categorization follows. The poster images for multiple multi-genre films are included in the dataset that we will be using. A folder with all those images in it is created.

1. First, import all the necessary Python packages.

```
import keras
from keras.models import Sequential
from keras.layers import Dense, Dropout, Flatten
from keras.layers import Conv2D, MaxPooling2D
from keras.utils import to_categorical
from keras.preprocessing import image
import numpy as np
import pandas as pd
import matplotlib.pyplot as plt
from sklearn.model_selection import train_test_split
from tqdm import tqdm
%matplotlib inline
```

2. Now, load the.csv file of input images of film posters and preprocess it.

```
train = pd.read_csv('multi_label_train.csv')    # reading the csv file

train.head()      # printing first five rows of the file

train_image = []

for i in tqdm(range(train.shape[0])):

    img = image.load_img('Multi_Label_dataset/Images/'+train['Id'][i]+'.jpg',target_size=(400,400,3))

    img = image.img_to_array(img)

    img = img/255

    train_image.append(img)

X = np.array(train_image)
```

3 Plot any one of the poster images.

This movie has a single genre—Thriller. Next, we would need the true label or labels for each of these images in our model.

4. Let us investigate. As far as we are aware, thirty genres are feasible. Each image will have 30 targets, indicating whether or not the movie fits into that genre. Thus, there will be 0 or 1 for each of these 30 targets. The train file's ID and genre columns will be eliminated, and the remaining columns will be converted to an array, which will serve as the destination for our images:

```
y = np.array(train.drop(['Id', 'Genre'],axis=1))

y.shape
```

5. Let us now construct a validation set to assist in evaluating our model's performance on hypothetical data. 10% of the images will be arbitrarily selected to serve as our validation set.

```
X_train, X_test, y_train, y_test = train_test_split(X, y, test_size=0.1, random_state = 42)
```

6. Designing our model's architecture is the next stage.

```
model = Sequential()
model.add(Conv2D(filters=16, kernel_size=(5, 5), activation="relu", input_shape=(400,400,3)))
model.add(MaxPooling2D(pool_size=(2, 2)))
model.add(Dropout(0.25))
model.add(Conv2D(filters=32, kernel_size=(5, 5), activation='relu'))
model.add(MaxPooling2D(pool_size=(2, 2)))
model.add(Dropout(0.25))
model.add(Conv2D(filters=64, kernel_size=(5, 5), activation="relu"))
model.add(MaxPooling2D(pool_size=(2, 2)))
model.add(Dropout(0.25))
model.add(Conv2D(filters=64, kernel_size=(5, 5), activation='relu'))
model.add(MaxPooling2D(pool_size=(2, 2)))
model.add(Dropout(0.25))
model.add(Flatten())
model.add(Dense(128, activation='relu'))
model.add(Dropout(0.5))
model.add(Dense(64, activation='relu'))
model.add(Dropout(0.5))
model.add(Dense(25, activation='sigmoid'))
```

7. Assemble the model now. ADAM as the optimizer and binary_crossentropy as the loss function are applied.

```
model.compile(optimizer='adam', loss='binary_crossentropy', metrics=['accuracy'])
```

8. To verify the model's performance, we have created validation data previously, and the model is also trained for ten epochs. We have seen that the accuracy of the model is almost 91%.

```
model.fit(X_train, y_train, epochs=10, validation_data=(X_test, y_test), batch_size=64)
Epoch 6/20
6528/6528 [==============================] - 24s 4ms/step - loss: 0.2528 - acc: 0.9079 - val_loss: 0.2535 - val_acc: 0
.9061
Epoch 7/20
6528/6528 [==============================] - 24s 4ms/step - loss: 0.2505 - acc: 0.9088 - val_loss: 0.2547 - val_acc: 0
.9060
Epoch 8/20
6528/6528 [==============================] - 23s 4ms/step - loss: 0.2484 - acc: 0.9087 - val_loss: 0.2575 - val_acc: 0
.9061
Epoch 9/20
6528/6528 [==============================] - 23s 4ms/step - loss: 0.2467 - acc: 0.9088 - val_loss: 0.2537 - val_acc: 0
.9061
Epoch 10/20
3712/6528 [================>.............] - ETA: 9s - loss: 0.2457 - acc: 0.9091
```

9. These two posters used: Titanic and Jaws for testing. Now let us feed these posters to our model. These photos need to be preprocessed before we can make any predictions.

$$img \Leftarrow load\_image(Titantic\ and\ Jaws)$$

$$img \Leftarrow preprocess\ (img)$$

10. We will now use our trained model to predict the genre of these posters. We will use the best three predictions from the model, which will give us the likelihood for each genre.

```
classes = np.array(train.columns[2:])
proba = model.predict(img.reshape(1,400,400,3))
top_3 = np.argsort(proba[0])[:-4:-1]
for i in range(3):
    print("{}".format(classes[top_3[i]])+" ({:.3})".format(proba[0][top_3[i]]))
plt.imshow(img)
```

Output

Movie Poster-1

Thriller (0.473)
Action (0.37)
Romantic (0.19)
Drama (0.2)

Movie Poster-2

Thriller (0.21)
Action (0.15)
Romantic (0.45)
Drama (0.32)

4.7.2 Multi-objective Image Segmentation

In this section, image segmentation is done using U-Net architecture, and multi-objective optimization is addressed using a bespoke loss function that combines binary cross-entropy and means squared error.

1. First, import all the necessary Python packages.

```
# Import necessary libraries
import numpy as np
import tensorflow as tf
from tensorflow.keras.models import Model
from tensorflow.keras.layers import Input, Conv2D, MaxPooling2D, UpSampling2D, concatenate
from tensorflow.keras.optimizers import Adam
from tensorflow.keras.losses import binary_crossentropy, mean_squared_error
from sklearn.model_selection import train_test_split
```

2. Load and preprocess the image dataset.

```
def load_and_preprocess_data(image_paths, mask_paths):
    images = []
    masks = []
    for img_path, mask_path in zip(image_paths, mask_paths):
        image = load_image(img_path)  # Implement this function to load and preprocess
        mask = load_mask(mask_path)   # Implement this function to load and preprocess
        images.append(image)
        masks.append(mask)
    images = np.array(images)
    masks = np.array(masks)
    return train_test_split(images, masks, test_size=0.2, random_state=42)
```

3. Define U-Net model for training.

```
def unet_model(input_shape):
    inputs = Input(shape=input_shape)

    # Encoding path
    c1 = Conv2D(64, (3, 3), activation='relu', padding='same')(inputs)
    p1 = MaxPooling2D((2, 2))(c1)

    c2 = Conv2D(128, (3, 3), activation='relu', padding='same')(p1)
    p2 = MaxPooling2D((2, 2))(c2)

    c3 = Conv2D(256, (3, 3), activation='relu', padding='same')(p2)
    p3 = MaxPooling2D((2, 2))(c3)

    # Bottleneck
    c4 = Conv2D(512, (3, 3), activation='relu', padding='same')(p3)

    # Decoding path
    u5 = UpSampling2D((2, 2))(c4)
    concat5 = concatenate([u5, c3])
    c5 = Conv2D(256, (3, 3), activation='relu', padding='same')(concat5)

    u6 = UpSampling2D((2, 2))(c5)
    concat6 = concatenate([u6, c2])
    c6 = Conv2D(128, (3, 3), activation='relu', padding='same')(concat6)
u7 = UpSampling2D((2, 2))(c6)
concat7 = concatenate([u7, c1])
c7 = Conv2D(64, (3, 3), activation='relu', padding='same')(concat7)

outputs = Conv2D(1, (1, 1), activation='sigmoid')(c7)  # Output layer

model = Model(inputs=[inputs], outputs=[outputs])
return model
```

4. Define the loss function and compile the model.

```
def custom_loss(y_true, y_pred):
    bce_loss = binary_crossentropy(y_true, y_pred)
    mse_loss = mean_squared_error(y_true, y_pred)
    return bce_loss + mse_loss  # Combine both losses
```

```
def compile_model(model):
    model.compile(optimizer=Adam(learning_rate=1e-4), loss=custom_loss,
    return model
```

5. Train the model on the preprocessed dataset.

```
history=model.fit (train_images, train_masks, validation_data = (
val_images,val_masks),batch_size=batch_size,epochs=epochs)

return history
```

6. Now, load the test image dataset and evaluate the model.

```
def evaluate_model(model, test_images, test_masks):
    evaluation = model.evaluate(test_images, test_masks)
    print(f"Test Loss: {evaluation[0]}, Test Accuracy: {evaluation[1]}")
```

```
train_images, val_images, train_masks, val_masks = load_and_process_data (image_paths, mask_paths)

input_shape = train_images [0]. shape

model = unet_model (input_shape)

model = compile_model (model)

history = train_model (model, train_images, train_masks, val_images, val_masks)

test_images, test_masks = load_test_data ()

evaluate_model (model, test_images, test_masks)
```

7. End.

4.8 Conclusions

This chapter emphasizes how these sophisticated machine learning approaches are more complex and essential for real-world applications where judgments need to take into account numerous criteria or categories at once. It covers a range of specialized algorithms and assessment metrics, including Pareto-based and evolutionary methods, that are intended to manage these complications. Techniques for feature selection in multi-label and multi-objective situations are explained. The importance of these techniques is highlighted by showcasing real-world applications in domains, such as multi-label and multi-objective image classification and segmentation using SVM and CNNs. The request for more approachable frameworks and toolkits to facilitate the wider application of these advanced classification methods comes at the end of this chapter.

Bibliography

1. Emmerich MT, Deutz AH (2018) A tutorial on multiobjective optimization: fundamentals and evolutionary methods. Nat Comput 17:585–609
2. Deb K, Sindhya K, Hakanen J (2016) Multi-objective optimization. In: Decision sciences. CRC Press, Boca Raton, pp 161–200
3. Saidabad MY, Hassanzadeh H, Ebrahimi SHS, Khezri E, Rahimi MR, Trik M (2024) An efficient approach for multi-label classification based on an advanced Kernel-Based learning system. Intell Syst Appl 21:200332
4. Han M, Wu H, Chen Z, Li M, Zhang X (2023) A survey of multi-label classification based on supervised and semi-supervised learning. Int J Mach Learn Cybern 14(3):697–724

5. Liu L, Martín-Barragán B, Prieto FJ (2021) A projection multi-objective SVM method for multi-class classification. Comput Ind Eng 158:107425
6. Tatsumi K, Hayashida K, Higashi H, Tanino T (2007) Multi-objective multiclass support vector machine for pattern recognition. In: SICE annual conference 2007. IEEE, pp 1095–1098
7. Xue B, Fu W, Zhang M (2014) Multi-objective feature selection in classification: a differential evolution approach. In: Simulated evolution and learning: 10th international conference, SEAL 2014, Dunedin, New Zealand, 15–18 Dec 2014. Springer, Berlin, pp 516–528
8. Spolaôr N, Cherman EA, Monard MC, Lee HD (2013) A comparison of multi-label feature selection methods using the problem transformation approach. Electron Theor Comput Sci 292:135–151
9. Ganda D, Buch R (2018) A survey on multi label classification. Recent Trends Program Lang 5(1):19–23
10. Coulibaly S, Kamsu-Foguem B, Kamissoko D, Traore D (2022) Deep convolution neural network sharing for the multi-label image classification. Mach Learn Appl 10:100422
11. Chin-Wei B, Rajeswari M (2010) Multiobjective optimization approaches in image segmentation—the directions and challenges. Int J Adv Soft Comput Appl 2(1):40–64
12. Dey N (ed) (2024) Applied multi-objective optimization. Springer, Singapore
13. Bong CW, Rajeswari M (2011) Multi-objective nature-inspired clustering and classification techniques for image segmentation. Appl Soft Comput 11(4):3271–3282
14. Zhang J, Zhou A, Zhang G (2015) A classification and Pareto domination-based multiobjective evolutionary algorithm. In: 2015 IEEE Congress on Evolutionary Computation (CEC). IEEE, pp 2883–2890
15. Oliva D, Abd Elaziz M, Hinojosa S, Oliva D, Abd Elaziz M, Hinojosa S (2019) Image segmentation as a multiobjective optimization problem. In: Metaheuristic algorithms for image segmentation: theory and applications, pp 157–179
16. Zhao B, Xue Y, Xu B, Ma T, Liu J (2018) Multi-objective classification based on NSGA-II. Int J Comput Sci Math 9(6):539–546
17. Mohapatra P, Das KN, Roy S, Kumar R, Dey N (2020) A novel multi-objective competitive swarm optimization algorithm. Int J Appl Metaheuristic Comput (IJAMC) 11(4):114–129
18. Bandyopadhyay S, Pal SK, Aruna B (2004) Multiobjective GAs, quantitative indices, and pattern classification. In: IEEE Trans Syst Man Cybern Part B (Cybern) 34(5):2088–2099
19. Bandyopadhyay S, Maulik U, Mukhopadhyay A (2007) Multiobjective genetic clustering for pixel classification in remote sensing imagery. IEEE Trans Geosci Remote Sens 45(5):1506–1511
20. Chatterjee S, Sarkar S, Dey N, Sen S, Goto T, Debnath NC (2017) Water quality prediction: multi-objective genetic algorithm coupled artificial neural network based approach. In: 2017 IEEE 15th International conference on industrial informatics (INDIN). IEEE, pp 963–968

Chapter 5
Deep Learning-Inspired Multiclass and Multi-label Classifications

5.1 Overview

Deep learning-inspired multiclass and multi-label classifications involve employing neural network architectures and methodologies inspired by deep learning to handle tasks where instances can belong to multiple classes simultaneously or where there are more than two classes. Effective categorization in complex circumstances is made possible by these techniques, which make use of neural networks' ability to autonomously build hierarchical representations of input. They frequently use activation functions and loss functions designed specifically for multiclass or multi-label classification problems, in addition to structures such as transformer models, recurrent neural networks, or convolutional neural networks (CNNs). Deep learning methods are particularly good at identifying complex relationships and patterns in the data, which makes them suitable for making precise predictions in a variety of fields, including image identification, natural language processing, and speech recognition. Addressing the shortcomings of conventional machine learning techniques when handling challenging classification jobs involving many classes or labels is the driving force behind deep learning-inspired multiclass and multi-label classifications [1]. Deep learning methods, which take inspiration from the structure and operation of the human brain, allow data to be automatically analyzed to identify complex patterns and relationships, leading to more reliable and accurate classification. These techniques are able to deal with the inherent complexity and subtleties found in real-world datasets by utilizing neural network topologies such as transformer models, recurrent neural networks (RNNs), and convolutional neural networks (CNNs). In addition, their scalability with massive amounts of data and flexibility across multiple domains and data types makes them especially desirable for applications ranging from recommendation systems and image identification to natural language processing and medical diagnosis. Therefore, the search for increasingly advanced and potent instruments to meet the difficulties presented by contemporary data analysis tasks is what motivates the development of deep

S. Chakraborty and L. Dey, *Multi-objective, Multi-class and Multi-label Data Classification with Class Imbalance*, Springer Tracts in Nature-Inspired Computing,
https://doi.org/10.1007/978-981-97-9622-9_5

learning-inspired multiclass and multi-label classifications [2]. The goals of deep learning-inspired multiclass and multi-label classifications are to improve classification performance in situations where conventional machine learning techniques are inadequate in terms of scalability, interpretability, and accuracy. The main objective is to create models that can efficiently handle complicated classification tasks involving numerous classes or labels by utilizing the capabilities of neural network architectures. Enhancing the capacity to precisely recognize and categories occurrences in a variety of fields, including image recognition, natural language processing, and medical diagnosis, is one way to do this. These goals also involve developing methods for feature representation learning, which will make it possible to automatically extract relevant features from raw data and improve model performance. Moreover, efforts are focused on improving interpretability of the models so that end users and domain experts can comprehend and rely on the judgements made by the deep learning models. The main goal is to push the limits of multiclass and multi-label classification in order to create new avenues for data-driven insights and applications across a range of industries [3].

5.2 Multi-class Classification Using Multi-layer Perceptron (MLP)

5.2.1 Background of MLP

An advanced artificial neural network architecture called a multi-layer perceptron (MLP) is made up of several feedforward layers of nodes, or neurons, organized in a certain pattern. An input layer, one or more hidden layers, and an output layer make up its composition. A network of interconnected nodes is formed when every neuron in one layer is connected to every other neuron in the layer below it. MLPs are often employed for tasks like pattern recognition, regression, and classification because they can learn intricate nonlinear correlations in data. The perceptron was originally designed to be an image recognition machine, even though it is now often known as an algorithm [4]. Its ability to see, and recognize images—a function similar to that of a human—gives it its name. The basic architecture of a typical MLP is given in Fig. 5.1.

A. **Input Layer**: Feature vectors are commonly used to represent input data that is received by this layer. In Fig. 5.1, X_1, X_2, ..., X_n are represented as input neurons.
B. **Hidden Layers**: Using weighted connections (W_i) and activation functions (f), hidden layers modify the incoming data. The hyperparameters that can be changed depending on the task's difficulty are the quantity of hidden layers and the number of neurons in each layer. In Fig. 5.1, hidden neurons are shown in the middle where different kinds of activations functions (Rectified linear unit

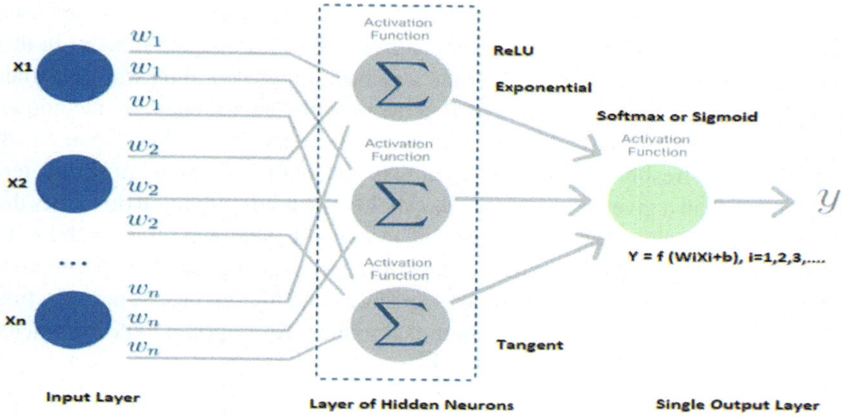

Fig. 5.1 A typical feedforward multi-layer perceptron

(ReLU), Tangent, exponential, etc.) can be used. By incorporating a constant to the input, or the specified bias, you can adjust the activation function using the Bias (*b*).

C. **Output Layer**: Final predictions or outputs are generated by the output layer. Depending on the type of job, such as binary classification (one neuron) or multiclass classification (many neurons), the output layer's neuron count differs. The last layer of Fig. 5.1 is the output layer where the final output is calculated using softmax or Sigmoid activation function.

$$Y = \sum f(W_i X_i + b)$$

Figure 5.1 is an example of a feedforward MLP because, like with the perceptron, inputs are added to the initial weights as a weighted sum and then run through the activation function. The propagation of each linear combination to the following layer is the distinction, though. The method would not be able to learn the weights that optimize the cost function if it just calculated the weighted sums for every neuron, transmitted findings to the output layer, and stopped there. Learning would not occur if the algorithm just performed one iteration. Backpropagation is useful in this situation. The multi-layer perceptron uses backpropagation as a learning technique to iteratively change the network's weights in an effort to minimize the cost function. Backpropagation needs one hard criterion in order to function effectively. Both the threshold function, such as ReLU, and the function that mixes inputs and weights in a neuron, such as the weighted sum, must be differentiable. Since the optimization function employed in multi-layer perceptrons is usually Gradient Descent, these functions must have a limited derivative. The gradient of the Mean Squared Error is calculated across all input and output pairs in each iteration after the weighted sums have been passed through all layers. The gradient value is then used to update the weights of the first hidden layer in order to propagate it back. In this manner, the weights are propagated all the way back to the neural network's initialization! Figure 5.2 shows a typical MLP network with backpropagation technique. This process continues until all input–output pairs have a converged gradient, which indicates that the freshly computed gradient has not changed from the previous iteration by more than a given convergence threshold. The below equation describes the one iteration of gradient descent method. MLPs are widely known for their ability to accurately depict complex relationships in data, but they can be sensitive to a number of hyperparameters, such as the number of neurons and hidden layers, the selection of activation functions, and regularization techniques. Appropriate hyperparameter tuning is essential for MLP performance [5].

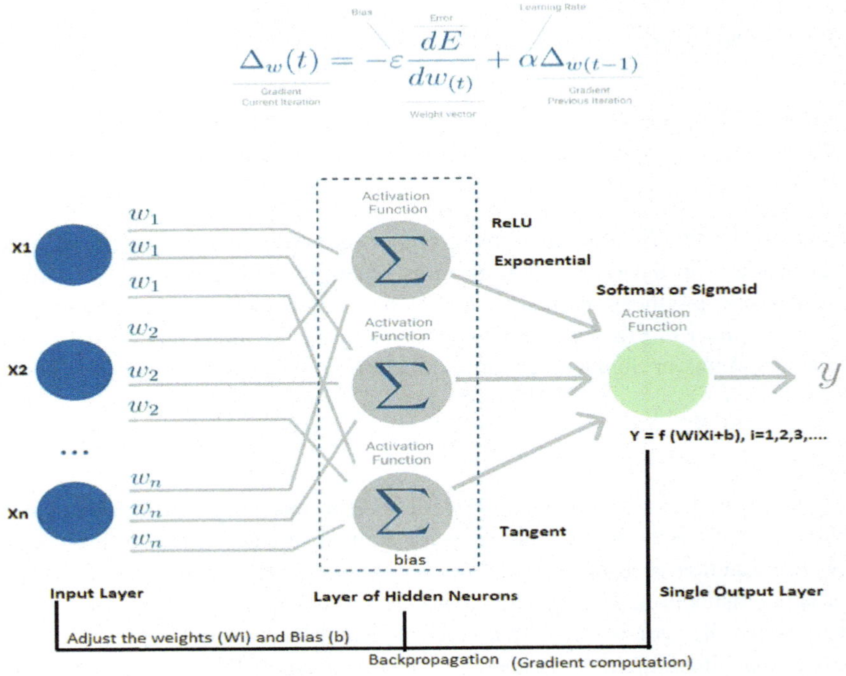

Fig. 5.2 A typical multi-layer perceptron with backpropagation

5.2.2 *Multi-class Classification Using MLP*

Dividing incoming data into one of the multiple predefined classes is the aim of a typical machine learning problem called multi-class classification utilizing a multi-layer perceptron (MLP). This is a step-by-step tutorial on utilizing MLPs to construct a multi-class classification model [6].

Step-1: Data Gathering: Compile and load a labeled dataset with each data point falling into a different class

```
LOAD dataset into train_data, train_labels, test_data, test_labels

# Flatten the input data (if using image data)

FOR each image in train_data:
        FLATTEN image

FOR each image in test_data:
        FLATTEN image

#Normalize the data,

        NORMALIZE train_data and test_data to range [0, 1]

#One-hot encode the labels

        ENCODE train_labels and test_labels as one-hot vectors
```

Step-2: Define the Model Architecture

```
INITIALIZE a sequential model

# Add input layer

        ADD dense layer with units=512, activation='relu',
input_shape=(input_dim,)

# Add hidden layers
        ADD dense layer with units=256, activation='relu'
        ADD dropout layer with rate=0.2 (optional, for regularization)
        ADD dense layer with units=128, activation='relu'
        ADD dropout layer with rate=0.2 (optional, for regularization)

# Add output layer
        ADD dense layer with units=num_classes, activation='softmax'
```

Step-3: Compile the Model

COMPILE model with optimizer='adam', loss='categorical_crossentropy', metrics=['accuracy']

Step-4: Train the Model

TRAIN model on train_data and train_labels with epochs=20, batch_size=128, validation_split=0.2

Step-5: Evaluate the Model

EVALUATE model on test_data and test_labels
PRINT test accuracy

Step-6: Fine-tune and Save the Model

ADJUST hyperparameters and re-train model as necessary

Important Aspects

Regularization: To avoid overfitting, use L2 or dropout regularization.

Batch Normalization: To stabilize and speed up training, normalize activations.

Early Stopping: Keep an eye on validation loss and halt training as soon as it ceases to advance.

5.2.3 Benefits of Multi-layer Perceptron (MLP) Classification

- **Nonlinearity Handling**: Because MLPs can model intricate, nonlinear relationships between features and target classes, they perform exceptionally well in a wide range of classification tasks.
- **Scalability**: Large MLPs can now be trained on high volume datasets, which makes them valuable in a range of applications. This is made possible by advancements in technology and libraries (like TensorFlow and PyTorch).
- **Feature Learning**: Significant feature engineering is not as required because MLPs automatically extract significant features from the data.
- **Parallel Processing**: Parallelizing MLP training and inference can expedite execution times on modern hardware.

- **Disadvantages of MLP classification**

There are various drawbacks when employing a multi-layer perceptron (MLP) for multi-class categorization. Preprocessing and feature engineering for MLPs are usually quite difficult and time-consuming. In addition, compared to more specialized models like Convolutional Neural Networks (CNNs) or Recurrent Neural Networks (RNNs), they struggle to handle high-dimensional data, like images or sounds. MLPs require rigorous regularization and validation techniques since they are prone to overfitting, particularly when there are a lot of classes and little training data. Additionally, they frequently need extensive hyperparameter adjustment to attain optimal performance and can struggle with scalability and computing efficiency, especially for very big datasets [4, 5].

5.3 Multi-class Classification Through Convolutional Neural Networks (CNN)

5.3.1 Background of CNN

A subclass of deep neural networks called convolutional neural networks, or CNNs, are frequently employed to analyze visual input. They work especially well for applications involving grid-like data, such object detection and image categorization. Figure 5.3 provides a typical architecture of convolutional neural networks (CNN) which is useful for image classifications task [7].

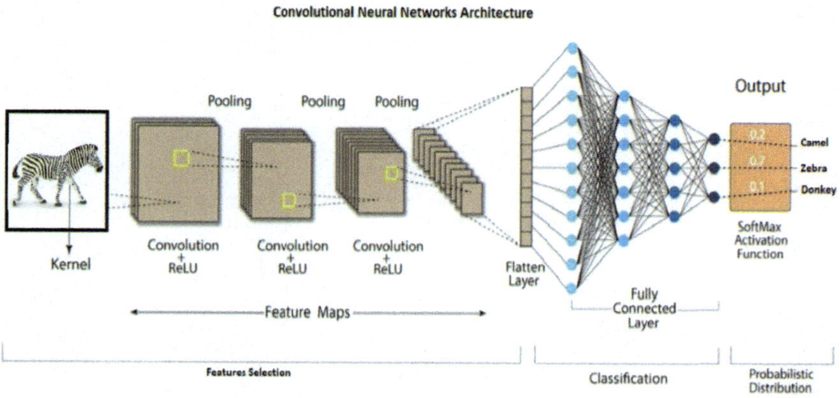

Fig. 5.3 A typical CNN architecture for image classification

The following describes the essential elements and functions of CNNs.

1. **Convolution Phase**

- **Filters/Kernel**: Tiny matrices that move over the input picture to identify characteristics like patterns, textures, and edges.
- **Convolution Operation**: The method of creating a feature map by applying the filter to the input. The filter performs element-wise multiplication and summing when convolved with the input.
- **Stride**: How many pixels the filter travels in a single step. The filter travels one pixel at a time when its stride value is 1.
- **Padding** is the process of enlarging the input image by pixels. There are two types of padding: "same" (where the output has the same dimensions as the input) and "valid," which involves no padding at all. Figure 5.4 shows a typical convolution operation on an image data.

2. Rectified Linear Unit, or ReLU, is a popular **activation function** in CNNs that adds nonlinearity to the model. ReLU sets the feature map's negative values all to zero. The size of feature map is (input – k) + 1, where input is the dimension of input image and k is the dimension of the kernel image.
3. **Pooling Phase**

- **Max Pooling:** By selecting the maximum value in a given window (such as 2×2) for each feature map, a technique known as max pooling reduces the spatial dimensions (width and height) of the feature maps.
- **Average Pooling:** This method is comparable to max pooling, except it uses the average value.

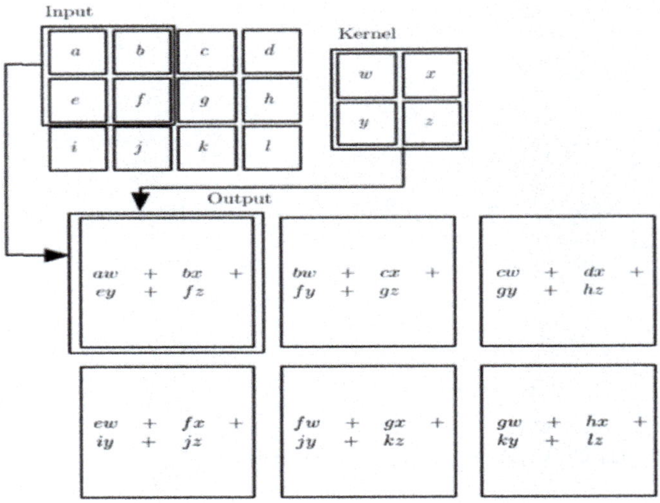

Fig. 5.4 A typical Convolution operation on an image

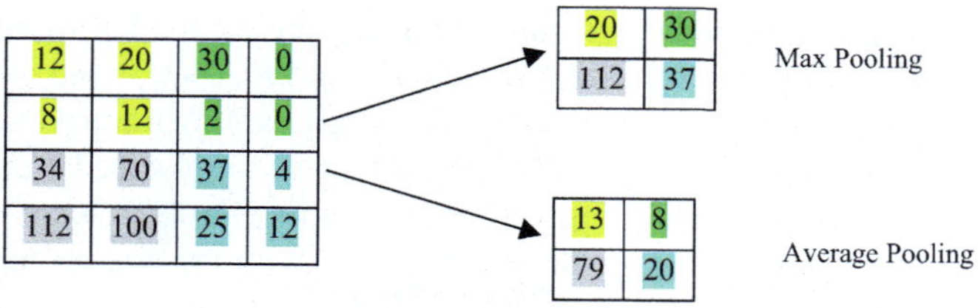

Fig. 5.5 Max and Average pooling operations on a sample image matrix

- By controlling overfitting and making the network invariant to tiny translations, pooling assists in lowering the computational effort. Figure 5.5 shows a max pooling and average pooling operations on a sample image.

4. The convolutional and pooling layers are preceded by the **fully connected** layers. They are employed to create predictions by combining the features that were extracted by the earlier layers. Every neuron in the layer above it is coupled to every other neuron in a fully connected layer.
5. **Flattening:** The feature maps' 2D matrix is flattened to provide a 1D vector that may be input into a fully linked layer. Figure 5.6 shows the use of ReLU activation function and flatten view of the data.
6. **Dropout:** A regularization method that, in order to avoid overfitting, randomly switches off a portion of the neurons during training.

There are various types of CNN networks are available, such as, LeNet, VGGNet, ResNet, AlexNet, GoogleNet, and ImageNet [8].

Classifying images into one of multiple predetermined classes is a typical problem in computer vision, where Convolutional Neural Networks (CNNs) are used for multi-class classification. Figure 5.3 provides a typical CNN architecture to classify a "Zebra" image with high probability. CNN networks are capable of performing one of the most fundamental yet effective computer vision tasks: multiclass image classification. We have two options for classification: We can download pre-existing image datasets or create our own labeled dataset. If there are few images for each class, preprocessing methods could involve tasks like image augmentation to add variance

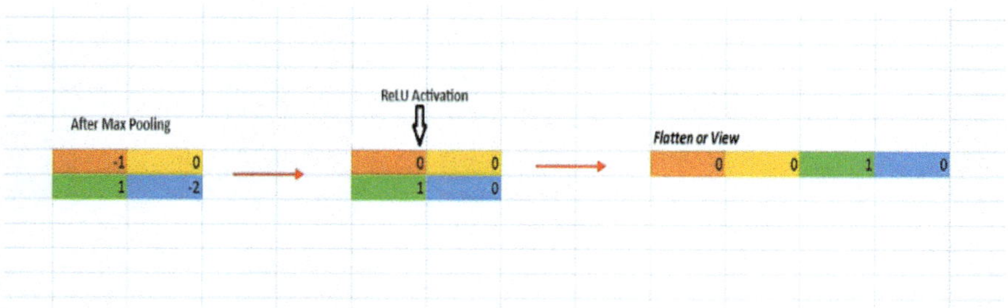

Fig. 5.6 Application of ReLU activation function and Flatten process

to the image collection. We have two options for training the model: we may utilize an existing backbone architecture, such as VGG16 and ResNet, or we can create a model architecture from scratch using any of the Deep Learning frameworks, such as TensorFlow or PyTorch. The benefit of the latter is that we simply need to concentrate on optimizing the model or modifying the final one or two levels in accordance with our use case, rather than having to start from scratch with the architecture. This is the context in which transfer learning is useful [9, 10].

5.4 Multi-class Classification Using Recurrent Neural Networks

One type of neural network that performs especially well with sequential data is called a recurrent neural network (RNN). They are an effective tool for tasks like language modeling, speech recognition, and time-series prediction since they are made to detect patterns and dependencies in time-series or sequential data. RNN is also used in stock market analysis, forecasting, natural language processing, image captioning, and so on.

5.4.1 Principal Characteristics of RNNs Sequential Processing

RNNs handle input sequences one-step at a time while preserving a hidden state that records data from earlier stages.

Memory: The network can remember information from earlier in the sequence thanks to the hidden state, which serves as a kind of memory. This is important for comprehending context in data (Fig. 5.7).

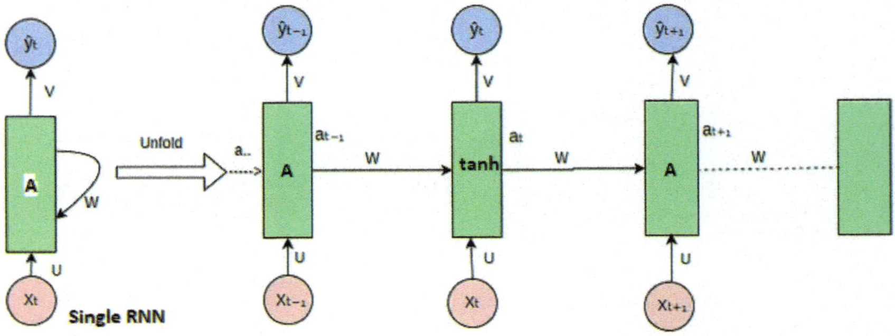

Fig. 5.7 A general RNN architecture

By progressively reading the data from left to right and updating the hidden state at every time step, the RNN accepts an input vector (X) and outputs a vector (Y). All time steps have the same set of parameters. This indicates that the network as a whole uses the same set of parameters, denoted by U, V, and W. W stands for the weight connected to the connection across hidden layers, V for the connection from hidden layer h to output layer y, and U for the weight parameter controlling the connection from input layer X to the hidden layer h. By keeping the knowledge from the previous input in its current hidden state, this parameter sharing enables the RNN to handle sequential data more quickly and accurately while capturing temporal dependencies [11, 12].

5.4.2 LSTM Architecture

A RNN faces problem during its backpropagation phase. Gradients may grow excessively big, causing the expanding gradient problem, or too little, causing the vanishing gradient problem, as they propagate backward through time during backpropagation. The problem with vanishing gradients is that they might get too small, making it difficult for the network to successfully identify long-term dependencies. Long short-term memory (LSTM) and gated recurrent unit (GRU) were introduced as a solution of this problem. LSTM is a special kind of RNN. Unlike RNN, LSTM is mainly designed to avoid the long-term dependency problem. A basic LSTM cell state architecture is shown in Fig. 5.8. One LSTM cell contains five parts,

1. A simple RNN cell.
2. Cell state—LTM

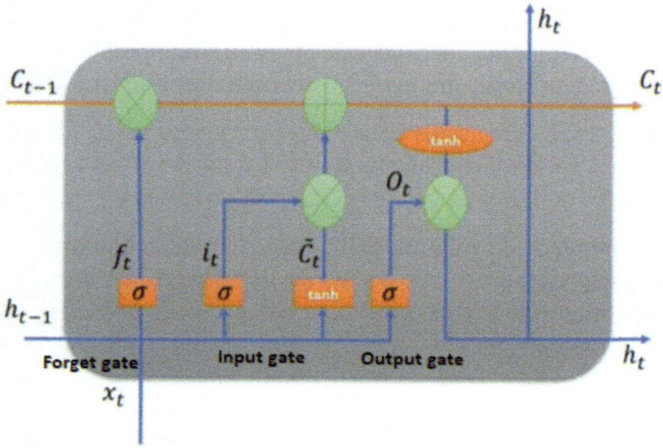

Fig. 5.8 A typical LSTM cell state

3. Forget gate.
4. Input gate.
5. Output gate.

$$
\begin{aligned}
f_t &= \sigma\left(x_t W_f + h_{t-1} U_f + b_f\right), \\
i_t &= \sigma(x_t W_t + h_{t-1} U_t + b_t), \\
o_t &= \sigma(x_t W_o + h_{t-1} U_o + b_o), \\
C_{t'} &= \tan\left[(x_t W_C + h_{t-1} U_c + b_c)\right], \\
C_t &= \sigma(f_t \cdot C_{t-1} + i_t \cdot C_{t'}), \\
h_t &= \tanh(C_t) \times o_t
\end{aligned}
\tag{5.1}
$$

An enhanced LSTM network is GRU. It combines the forgotten port and the incoming port into one updated port. GRU fully uses just integration to expose its memory content [11].

5.4.3 Multiclass Text Reviews Classification

Classifying an input sequence into one of multiple predetermined categories is the aim of multi-class classification [11]. Typically, RNNs are applied as follows to this task:

Input Representation: A series of vectors typically makes up the input. For example, in text categorization, word embeddings (e.g., Word2Vec, GloVe) are frequently used to represent words or characters as vectors.

Network Architecture: The sequence is processed by an RNN, such as a gated recurrent unit (GRU) or long short-term memory (LSTM). Because these variations may better handle long-range dependencies by mitigating problems like vanishing and exploding gradients, they are frequently chosen over standard RNNs.

Output Layer: A fully connected (dense) layer receives the last hidden state (or a series of hidden states), which is then followed by a softmax activation function. A probability distribution over the possible classes is produced by the softmax layer.

Training: An optimizer like Adam or SGD is used in conjunction with a suitable loss function, like categorical cross-entropy, to train the network. As the model is trained, it gains the ability to modify its weights in order to reduce the loss and increase classification accuracy.

Case Study

The achievement of LSTM and GRU in terms of accuracy for categorizing text reviews provided for movies on a scale of one to ten is the main subject of this case study. Numerous entertainment businesses, such as Netflix and Amazon, have the ability to utilize sentiment analysis of movie reviews to recommend films with better ratings to their viewers [13].

In this case study, the dataset of one lakh of movie review samples is collected from IMDB. The experimented dataset has eight distinct classes (1, 2, 3, 4, 7, 8, 9, 10) of reviews. The sample of the final experimented dataset is given in Table 5.1.

The text reviews that are the subject of this study are produced by people using natural language; they may or may not be structured. Thus, preprocessing text reviews is necessary. Several filters, such as converting text to lowercase and deleting HTML tags, are applied to all of the text reviews to make them consistent. Internet slang, modern spellings, such apostrophes, and word forms, to mention a few, and informal language are commonly employed in online reviews. Therefore, for accurate data retrieval, such terms need to be reviewed and stemmed. Porter stemming is a method for removing suffixes like as "ing," "ly," "s," and so on in order to get at the root of a word. Porter Stemmer in the Python language's nltk (Natural Language Toolkit) module is utilized in this experiment on the dataset under study. Therefore, the unnecessary stop words have been removed to make the dataset suitable for information retrieval and text mining using LSTM and GRU. To achieve this, at first tokenization of the data has been done. Tokenize text reviews for movies using the 5000 most common words. After tokenization, the words are vectorized. However, 500 sequence padding is used to make sure every review is the same length. The accuracy results of this case study is described in Table 5.2.

Table 5.1 Dataset of sample movie reviews

Sentiment	Movie review
1	Simply a stilted copy of the infinitely superior…
2	The first movie was fantastic, and the second one was…
3	A somewhat inadequate Firestarter clone that appears more….
7	Alright, so it might not be the highly regarded….
10	This has to be the greatest computer ever made……

Table 5.2 Accuracy results

Model	Preprocessing	Training accuracy (%)	Validation accuracy (%)	Test accuracy (%)
LSTM	Porter stemming	97	78	77
LSTM	Stop words	97	77	76
LSTM	Basic	97	77	77
GRU	Porter stemming	96	78	20
GRU	Stop words	95	77	76
GRU	Basic	96	77	76

5.5 Multi-class Classification Using Deep Generative Networks

Deep generative models, which can describe complex data distributions and produce new data samples, have gained popularity. Examples of these models are variational autoencoders (VAEs), diffusion model, transformer model, and generative adversarial networks (GANs). Multi-class classification challenges can also benefit from the adaptation of these models.

For multi-class classification, Generative Adversarial Networks (GANs) can be modified by incorporating an additional classifier into the discriminator [14]. The generator in this configuration generates synthetic data samples from random noise, maybe conditioned on class labels, and the discriminator predicts class labels in addition to differentiating between produced and real samples. Training teaches the discriminator to accurately categorize both real and created samples while simultaneously teaching the generator to create realistic samples that can trick the discriminator. By utilizing the rich, diverse data produced by the generator, this adversarial process improves the discriminator's performance in multi-class classification problems and produces a strong classifier with good cross-class generalization capabilities. By adding a classifier to the latent space framework, variational autoencoders (VAEs) can be modified for multi-class classification.

VAEs are made up of an encoder that converts input data into a latent space that is represented by a distribution and a decoder that uses this latent representation to recreate the data. In order to perform classification, an additional classifier network is added to the latent representations in order to predict the class labels that condition the latent space on. While the classifier simultaneously learns to correctly anticipate class labels based on these latent representations, the VAE learns to encode the data into a meaningful latent space that preserves class information during training. By providing a structured and informative latent space, this method makes use of the VAE's capacity to capture complex data distributions, improving the classifier's performance on multi-class classification problems [15].

Diffusion models can be used for multi-class classification by taking use of their capacity to simulate intricate data distributions using iterative noise reduction procedures. Diffusion models were initially created for generative tasks. These models capture the underlying structure of the data by progressively converting a basic noise distribution into a data distribution. A diffusion model can be trained to predict both the class probabilities and the denoised data at each step in the classification process. This is accomplished by training the model to recognize discriminative characteristics that distinguish between classes during the reverse diffusion phase by conditioning the denoising process on class labels. By combining generating abilities with class-conditioned discriminative tasks, diffusion models can offer a strong foundation for multi-class classification and produce accurate class predictions [16].

Transformer models, originally developed for sequence-to-sequence tasks in natural language processing, are highly effective for multi-class classification due to their self-attention mechanism, which captures long-range dependencies and contextual relationships within the data. In a multi-class classification task, a transformer model can be adapted by using its encoder architecture to process the input sequence, whether text, image patches, or other structured data, and generate a contextualized representation of the input. This representation is then passed through a classifier, typically a fully connected layer followed by a softmax activation function, to output probability distributions over the multiple classes. The attention mechanism allows the transformer to weigh the importance of different parts of the input, leading to improved classification performance by focusing on the most relevant features. This approach has proven highly effective across various domains, including text classification, image recognition, and other structured data classification tasks.

Transfer learning is used by several generative AI models, including GPT (generative pre-trained transformer). One area of machine learning research that needs attention is transfer learning. It retains the information learned from solving a particular problem and uses it to solve another that is similar but unrelated. Transfer learning, a method in deep learning, involves training a neural network model on an example problem before applying it to the actual problem at hand. One benefit of transfer learning is that it can reduce generalization mistakes and shorten the training period for a learning model. To suit our needs, we can employ a pre-trained model that was trained on different datasets, such as ImageNet, and adjust the final layers. When training a model from scratch, we might be able to save time, energy, and resources in this situation. Due to intensive training, these models have a vast amount of image structure and data about previously learned images [17].

Transformer-based architectures and other generative AI (GENAI) models can be very successful in multi-class classification challenges. The bidirectional encoder representations from transformers (BERT) is one such potent GENAI model that can be adjusted for a variety of categorization issues. Suppose, we describe a BERT model step-by-step for multiclass text classification [18].

Step-1: Collect text label data with multi-group categories.
Step-2: Transfer text labels into numerical values.
Step-3: Dataset needs to be splitted into train and test sets.
Step-4: Tokenize text using BERT's tokenizer, making sure each token is the same length by padding and truncating.
Step-5: After loading BERT, compile the model using accuracy metrics, a sparse categorical cross-entropy loss, and an Adam optimizer.
Step-6: Validate on the test set after a few epochs of training the model.
Step-7: Predict on the test set, compute accuracy, and translate logits to class labels.

Because transformer-based models, such as BERT, can comprehend contextual relationships in text input, they are very useful for multi-class classification tasks. As this example shows, fine-tuning a pre-trained BERT model for particular classification tasks can result in excellent accuracy and resilient performance.

5.6 Literature Survey

This section of this chapter is essential as it provides a comprehensive overview of existing knowledge and research on multiclass and multi-label classification using deep learning models. It helps to identify gaps, inconsistencies, and the current state of understanding, thereby justifying the need for the current study. In this section, we discuss a collection of novel and intriguing works on multiclass classification utilizing various deep learning techniques, especially focusing on deep generative networks.

As with many medical applications [19], automatic skin issue diagnosis is difficult due to a lack of annotated datasets. Although deep learning methods have increased the accuracy of diagnoses, the availability of labeled data limits how successful they may be. While helpful, traditional data augmentation techniques fall short, generative adversarial networks (GANs) have demonstrated potential recently in producing artificial images to supplement training data. In order to overcome class imbalance, a work [14] introduces a deep generative adversarial network (DGAN) multi-class classifier that uses images from many Internet databases to produce skin issue images. One of the biggest challenges in training was stability. Using conventional data augmentation approaches, two CNN models based on ResNet50 and VGG16 were used to analyze the performance of the DGAN. For testing, both labeled and unlabeled data were utilized. A generative model can be used to create new data samples that are comparable to existing ones by concurrently capturing the joint probability of the input data and labels Prob(x_int, y_lbl). When using images as input data, for instance, each sample (image) comprises hundreds of dimensions (pixels), and the task of the generative model is to preserve the dependencies between pixels, such as the possibility that adjacent pixels may combine to form a recognizable object [20]. A typical GAN architecture is shown in Fig. 5.9. A novel technique called Generative OpenMax (G-OpenMax) [21] is presented to produce artificial samples from unidentified classes. In order to produce reasonable representations with regard to the known classes' domain, the synthetic samples are created using mixture distributions of known classes in latent space. It is applied to the difficult problem of multi-class open set classification. In order to produce believable and domain-adapted synthetic unknown samples, the training process is updated using conditional GAN and a combination of many previous known class distributions.

$$\min_{G}\max_{D} V(D,G) = E_{x\sim_{\text{data}(x)}}[\log D(x)] + E_{z\sim_{Pz(z)}}[\log(1 - D(G(z)))] \tag{5.2}$$

In the above equation, G (generator) uses transposed convolution processes to create images using a random Gaussian noise vector as an input. D (discriminator) has been trained to differentiate between generated false images ($G(z)$) and actual photos (x). Consider the following minimax game with the value function V (G, D) for the optimization of D and G. G is trained to reduce D's capacity to discriminate between actual and generated images, and D aims to increase the likelihood that real training images will be labeled as "real" and G's generated images as "fake." While

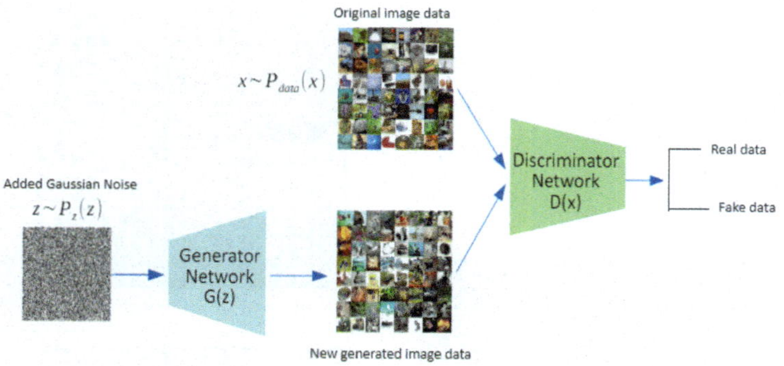

Fig. 5.9 A typical GAN architecture to create fake data and discriminate between real and fake data

the discriminator becomes stronger at accurately differentiating between real and created images, the generator gets better at producing more realistic images.

In the Generative OpenMax (G-OpenMax) [21] technique, the network Net^G is trained with $K + 1$ classes instead of training the classifier and Weibull model at the first stage by using known classes $c_{i,\ldots,K}$ only in parallel, where the other images used to represent the unknown class are taken from generator G. One additional dimension is output by the activation vector v^G in G-OpenMax to represent the $K + 1$ class, which is thought to be unknown. It is observed that dependent information regarding the relationships between classes is provided by the penultimate layer. Consequently, in feature space, the additional dimension can provide information on the MAV distance between known and unknown classes. In the second step, G-OpenMax uses Weibull CDF probability for score calibration, following the same steps as OpenMax. A fascinating piece of work to learn data representation from a small number of samples, MFC-GAN [22], has been proposed. By adding synthetic samples to the original data using MFC-GAN, it can manage datasets with unequal distributions of classes. Generated images are classified into various fake classes and actual images into real classes during the training phase. The labels of the original training data and the corresponding fake labels for the generated images are genuine labels. Each label was transformed into a n bit one-hot encoding vector, where n is the number of classes, in order to prepare the class labels. n zeros are padded to the right of the label encoding to produce a new representation for actual labels, which allows for false classes. Therefore, by padding n zeros to the left of the original label encoding, a corresponding fake class label Fc is created for each actual label RC. With a modified AC-GAN aim, MFC-GAN is trained. Maximizing the log-likelihood of categorizing fake samples into fake classes FC and real samples into real classes RC is the goal. The genuine labels are used to condition the MFC-GAN generator in order to produce class-specific samples. Figure 5.10 represents the results of MFC-GAN-generated images on MNIST images.

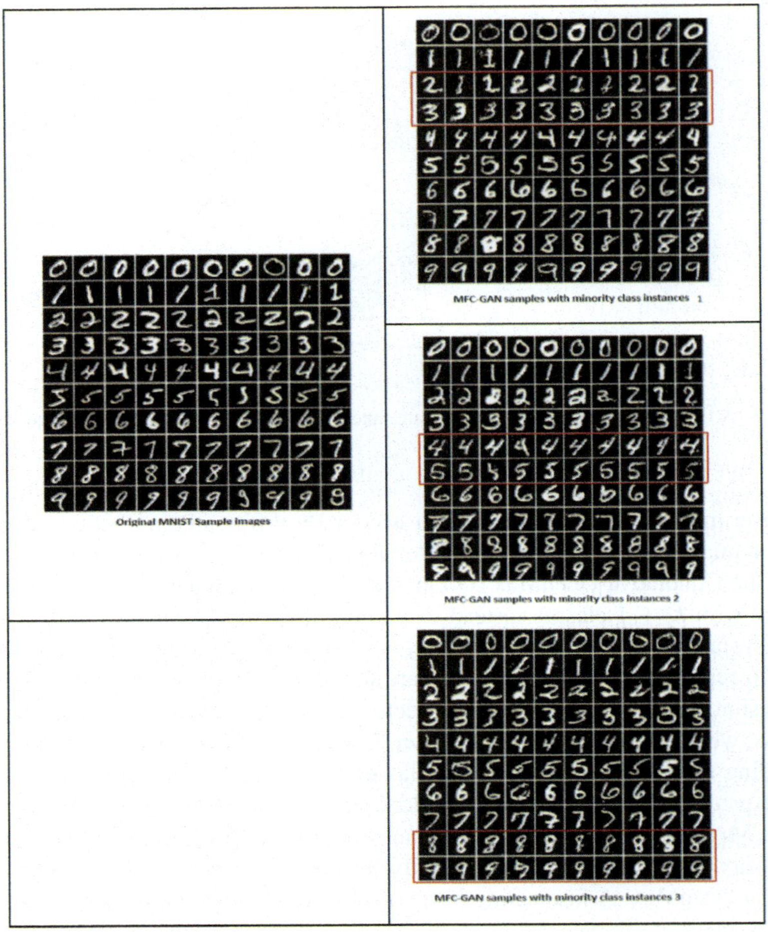

Fig. 5.10 Original images using MFC-GAN samples from MNIST dataset, with red-highlighted occurrences of minority classes

The majority of the work that has already been done uses deep generative models for single label learning tasks, despite the obvious benefits of a probabilistic interpretation of observed data. Although generative methods for multi-label classification (MLC) have been studied in the literature, most of the solutions now in use require fully or at least partly labeled training data. Stated differently, it is difficult to apply them to the solution of semi-supervised MLC or even weakly supervised MLC tasks. There is one research work [23] which deals with these kind of challenges of weakly supervised MLC (WS-MLC) tasks. A unified probabilistic framework is suggested to explain the multi-label data for WS-MLC using a deep sequential generative model. We offer a deep sequential classification model in our model that can be used for posterior inference approximation and prediction. It also create effective learning algorithms using variational inference that can be applied to WS-MLC with promising results. The goal of multi-label zero-shot learning is to categorize images into several previously unidentified categories for which no training data is supplied. A fusion approach is embedded with generative adversarial network (GAN) to do multi-label zero-shot learning [24]. This paper introduces various fusion algorithms for synthesizing multi-label features from their associated multi-label class embeddings at the attribute, feature, and cross-level. The primary goal of creating a generative multi-label zero-shot strategy is to create multi-label visual features that are semantically consistent by combining relevant class properties. There are two approaches to synthesis multi-label visual features. i) Integrating class-specific attribute embeddings at the generator's input to create a global image-level embedding vector is the first attribute-level fusion (ALF) technique. One such method is to separately synthesize the features derived from the class-specific embeddings and subsequently incorporate them into the visual feature space. This method is known as feature-level fusion, or FLF.

The alternative fusion strategy of this work, known as cross-level fusion (CLF), combines the benefit of label dependency of ALF and the class-specific discriminability of FLF. The CLF method takes into account the bi-level context (from ALF and FLF) and makes use of every individual-level feature. Because of this, individual-level features self-adapt to create enhanced synthesized features, which are combined to generate the CLF output. We examine the suggested multi-label feature generation CLF strategy for (generalized) zero-shot object identification in addition to multi-label zero-shot classification.

A major problem with GANs is that they may learn to generalize and make it more difficult to distinguish between two groups of images if they have similar properties. A modification to the conventional GAN training method (MCGAN) is presented in a research study [25], employing tiny sets of labeled data to enable better categorization in related classes of images inside a semi-supervised learning framework. The authors force G to refrain from generalizing in a way that could produce images that are comparable to those of other classes by utilizing labeled images that are currently accessible from both classes to train the MCGAN. The proposed GAN-based multi-class (MCGAN) classifier aims to discriminate between two classes of data ($C1$, $C2$), where little labeled data is available for class $C2$ while available labels are available for class $C1$. An authentic picture (labeled Real) and the generator's output (labeled Fake) are the inputs of a conventional GAN discriminator (DCGAN, AnoGAN, etc.). As a result, the generator is forced to discover how the images are distributed in the actual class. If the features of $C1$ and $C2$ photos are comparable, then training the GAN on C1 images may enable G to learn and generalize sufficiently, resulting in the generation of images that are similar to $C2$, hence impeding the categorization of the two classes ($C1$ vs. $C2$). In order to circumvent this difficulty, we provide labeled photos of $C2$ as a third input to the discriminator. Even though these are genuine photos from $C2$, we classify them as fake. This compels the generator to learn the properties of $C1$ rather than generalize to this comparable class ($C2$). The discriminator marks an image that G creates that might pass for one of $C2$'s images, and G reassesses its learning at such positions. The architecture of a typical MCGAN is represented in Fig. 5.11.

The variation score var(x) for an input image x is represented as,

$$\mathrm{Var}(x) = (1 - £) \times L_R(z) + £ \times L_D(z). \tag{5.3}$$

where the residual and discriminator losses, $L_R(z)$ and $L_D(z)$, respectively, enforce the visual and image characteristic similarity between the produced picture $G(z)$ and the true image x. The discriminator functions as an image encoder when the discriminator loss uses the output of an intermediary layer, $f(.)$, to record image attributes. The pixel-by-pixel disparity between picture x and $G(z)$ is known as residual loss.

$$\begin{aligned} L_R(z) &= \sum |x - G(z)|, \\ L_D(z) &= \sum |f(x) - f(G(z))|. \end{aligned} \tag{5.4}$$

Prior to determining Var(x) during the test, a point z_i must be identified via back-propagation, which attempts to produce an image that is as close to image x as feasible. z_i is found using a loss function that is based on the discriminator and residual losses that are described below in Eq. (5.5).

$$L(z_i) = (1 - £) \times L_R(z_i) + £ \times L_D(z_i). \tag{5.5}$$

The weighted total of the overall loss and variation score is adjusted by £.

A unique deep sequential generative model for semi-supervised multi-label learning from crowds is proposed in a work [26]. The true label latent vectors and the data instances latent features are two new latent variables in this model. Furthermore, these two latent variables are used to construct the labels of the crowds. A crowd's multi-labeled data can be fully utilized by the suggested deep sequential generative model in conjunction with unlabeled data. Additionally, it incorporates a

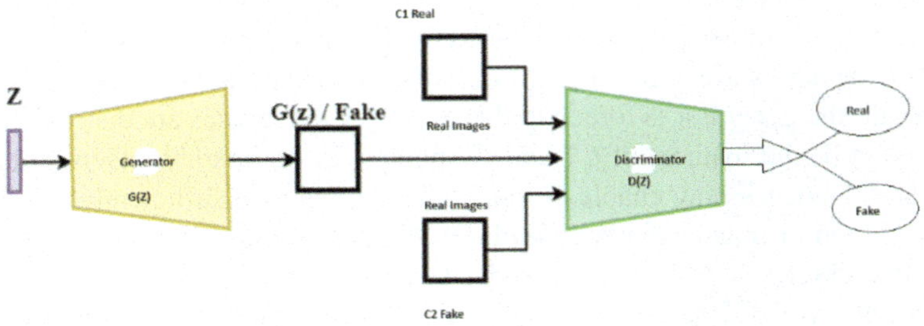

Fig. 5.11 Architecture of MCGAN

deep sequential inference model for learning ground truth labels and posterior inference. A novel probabilistic generative model is used in a research study to address the multi-label classification problem [27]. The model's goal is to depict the fusion of numerous active concepts in feature creation. The idea prediction problem was framed as probabilistic inference in a graphical model by associating and introducing exclusive latent random variables to express the contribution levels for the concepts with different representative basic feature vectors. For the suggested model, two-parameter learning strategies are presented: One is based on the expectation–maximization approach, which is a typical maximum likelihood learning method, and the other one focuses on maximizing the margin of the actual configuration's separation from the class boundary. Several multi-label image classification tasks show a significant improvement in performance using these learning algorithms. Each notion in this model has many representative base feature vectors associated with it, and the final feature vector is created by combining the base features in a significance-weight linear fashion. It is described by the below Eq. (5.6).

$$\mathbf{x} = \sum_{j=1}^{K} s_j \mathbf{b}_{z_j}^{j} + \mathbf{v}, \quad \text{where } \mathbf{v} \sim \mathcal{N}(0, \sigma^2 \mathbf{I}) \tag{5.6}$$

The feature vector **x** is thought to be constructed as a significance-weighted sum over the K concepts given the chosen base feature vectors ($\mathbf{b}_{z1}{}^1$, $\mathbf{b}_{z2}{}^2$, …, $b_{zK}{}^K$) and the accompanying significance levels (s_1, s_2, …, s_K). v is the added Gaussian spherical noise in the emission process [27].

Four distinct eye illnesses (multi-classes)—diabetic, cataract, glaucoma, and normal—are represented using ResNet50 and InceptionV3 models in a research study [28]. The Generative Adversarial Network (GAN) model is used to enrich the retina image dataset. It is essential to guarantee the fairness and impartiality of the dataset images. Once our ensemble process was first initialized with simple ImageNet data and then tweaked, it performed a fantastic job of learning from these datasets. The study examined the application of two well-known deep learning multi-class classification models, ResNet50 and InceptionV3, which were chosen for their superior architectural features, using this upgraded dataset as a basis.

ResNet50's residual learning architecture effectively handles the vanishing gradient problem by enabling the model to learn from a deep depth of layers. Conversely, the InceptionV3 inception approach extracted characteristics at multiple scales by combining different sized filters. Because lesions in retinal imaging can vary widely in size, this technique is particularly helpful for assessing disorders related to retinal imaging. The overall classification results is shown in Fig. 5.12.

An interesting research work [29] that converts multiclass image classification into multi-label classification using ResNet50, VGG16, and VGG19 models. Authors are working to enhance recognition of image tasks that can identify different labels that may be present in images with low-resolution, distracted backgrounds, to supplement the robust object detection models that are in great demand in the rapidly growing field of computer vision. Older or low-quality images can be upgraded from 32 × 32 pixels

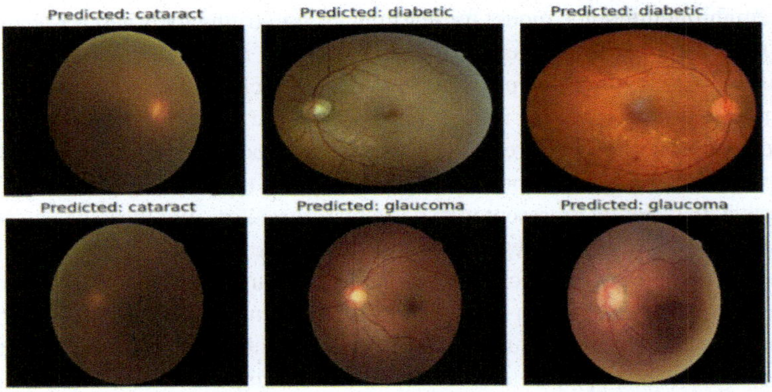

Fig. 5.12 Multiclass eye disorders classification results of ResNet50 and InceptionV3

to 227 × 227 pixels using ESRGAN (Enhanced Super-Resolution Generative Adversarial Networks) to increase their resolution and aesthetic appeal. It is an advanced deep learning model for tasks related to image super-resolution, that is, improving the resolution and quality of images. The model is trained on a large dataset of both types of image pairs in order to learn the mapping between high-resolution and low-resolution image pairs. The authors defined their custom labels in accordance with the modifications made to the dataset for multi-label image categorization. The classes in the dataset are redesigned and new labels are added. The purpose of augmenting image data is to train the model with these modified images, making it more resilient and better able to handle various changes and circumstances during inference. The aim of this work is to build a model that can effectively extract information from images and improve the robustness and precision of image categorization under these specific circumstances [29].

5.7 Hands-On Experience of Multiclass Classification Using GAN and Transfer Learning

Using a Generative Adversarial Network (GAN) for multiclass classification entails taking advantage of the network's ability to provide realistic data samples that can help train a classifier. Usually, there are two primary parts to the process: a discriminator that separates actual data from fake and a generator that creates artificial data that looks like the original dataset. These elements compete during training, resulting in the creation of data that is more realistic. Once trained, a classifier's performance on multiclass classification tasks can be enhanced by using the synthetic data produced by the GAN to improve the classifier's training. This method is particularly helpful in situations where there is a lack of labeled data because the GAN can contribute more high-quality synthetic examples to the dataset. The pseudocode of multiclass classification on a public dataset using GAN model is described below.

Pseudocode

```
# Include all significant libraries in Python

importing torch and torch neuralnetworks
import torch optim for parameter adjusting during the training phase
importing DataLoader to load the dataset
import datasets, transforms using torchvision to perform typical image transformation

# Generator Model Definition

Define Generator class (inherits nn.Module):
   function init():
      call super(Generator, self).__init__()

      # Define the neural network layers in sequence
      self.model -> Sequential(
         Linear(input_size=100, output_size=256), # Input layer
         ReLU(), # Activation function
         Linear(input_size=256, output_size=512), # Hidden layer 1
         ReLU(), # Activation function
         Linear(input_size=512, output_size=1024), # Hidden layer 2
         ReLU(), # Activation function
         Linear(input_size=1024, output_size=3*32*32), # Output layer
         tanh() # Activation function to scale outputs between -1 and 1
      )

   function forward(x):
      # Forward pass through the network
      output -> self.model(x)

class Discriminator (inherits nn.Module):
   function init():
      call super(Discriminator, self).__init__()

      # Define the neural network layers in sequence
      self.model -> Sequential(
         Linear(input_size=3*32*32, output_size=1024), # Input layer
         LeakyReLU(negative_slope=0.2), # Activation function
         Linear(input_size=1024, output_size=512), # Hidden layer 1
```

```
        LeakyReLU(negative_slope=0.2),  # Activation function
        Linear(input_size=512, output_size=256),  # Hidden layer 2
        LeakyReLU(negative_slope=0.2),  # Activation function
        Linear(input_size=256, output_size=1),  # Output layer
        Sigmoid()  # Activation function to scale outputs between 0 and 1
    )

  function forward(x):
    # Reshape the input to have dimensions (batch_size, 3*32*32)
    x -> x.view(-1, 3*32*32)

    # Forward pass through the network
    output -> self.model(x)
     return output

Classifier class (inherits nn.Module):
  function init():
     call super(Classifier, self).__init__()

     # Define the neural network layers in sequence
     self.model -> Sequential(
        Linear(input_size=3*32*32, output_size=512),  # Input layer
        ReLU(),  # Activation function
        Linear(input_size=512, output_size=256),  # Hidden layer
        ReLU(),  # Activation function
        Linear(input_size=256, output_size=10)  # Output layer (assuming 10 classes
for CIFAR-10)
     )

  function forward(x):
     # Reshape the input to have dimensions (batch_size, 3*32*32)
     x -> x.view(-1, 3*32*32)

     # Forward pass through the network
     output -> self.model(x)
     return output

# Load the CIFAR-10 dataset

Transform, resize and normalize the dataset to match the generator output size
```

```
Split into train and test datasets

train_loader <- dataloader(train_dataset)
test_loader <- dataloader(test_dataset)

# Initialize the models
generator -> Generator()
discriminator -> Discriminator()
classifier -> Classifier()

# Define loss functions and optimizers
criterion_GAN -> nn.CELoss()
criterion_class -> nn.CrossEntropyLoss()
optimizer_G -> Adam(generator.parameters(), learning=0.0002, betas=(0.5, 0.999))
optimizer_D  ->  Adam(discriminator.parameters(),  learning=0.0002,  betas=(0.5,
0.999))
optimizer_C -> Adam(classifier.parameters(), lr=0.0002)

# Training loop
for epoch -> 50
   train() -> generator
   train() -> discriminator
   train() -> classifier

   for each batch (images, labels) in train_loader:
      batch_size  ->  size(images, 0)

      # Create labels for real and fake data
      real_labels  ->  ones(batch_size, 1)
      fake_labels  ->  zeros(batch_size, 1)

      # Train Discriminator
      optimizer_D.zero_grad()

      # Real images
      outputs -> discriminator(images)
      loss_real  ->  criterion_GAN(outputs, real_labels)

      # Fake images
      noise  ->  randn(batch_size, 100)
      fake_images  ->  generator(noise)
```

```
        outputs -> discriminator(fake_images.detach())
        loss_fake -> criterion_GAN(outputs, fake_labels)

        # Backprop and optimize
        loss_D -> loss_real + loss_fake
        backward(loss_D)
        optimizer_D.step()

        # Train Generator
        optimizer_G.zero_grad()

        # Generate fake images
        outputs -> discriminator(fake_images)
        loss_G -> criterion_GAN(outputs, real_labels)

        # Backprop and optimize
        loss_G.backward()
        optimizer_G.step()

        # Train Classifier
        optimizer_C.zero_grad()

        outputs -> classifier(images)
        loss_C -> criterion_class(outputs, labels)

        # Backprop and optimize
        loss_C.backward()
        optimizer_C.step()

# Evaluate the classifier
classifier.eval()
correct -> 0
total -> 0
for each (images, labels) in test_loader:
    outputs -> classifier(images)
    _, predicted -> max(outputs.data, 1)
    total + -> size(labels, 0)
    correct + -> sum(predicted .equals( labels)).item()

print (Accuracy of the classifier on the test images)
```

Discussion of Pseudocode

Fig. 5.13 Sample training data of CIFAR-10 dataset

- **Generator Model**: The generator creates realistic visuals by feeding it random noise.
- **Discriminator Model**: Real or fraudulent images are classified by the discriminator.
- **Classifier Model**: Real images are classified into one of the ten CIFAR-10 classes using a straightforward feedforward neural network.
- **Loss Functions**: Binary cross-entropy is used to define the GAN loss. A cross-entropy loss is used by the classifier.
- **Data loading**: PyTorch is used to load and preprocess the CIFAR-10 public dataset. The training data samples are given in Fig. 5.13.
- **Training Loop**: To raise the caliber of images produced, the discriminator and generator are trained alternately. Real images are used to train the classifier.
- **Evaluation**: The classifier is assessed on the test set at the end of each epoch, and the test accuracy and loss are reported.

1. **Results on execution of GAN on CIFAR-10 Dataset**

The above code has been executed over Google Colab GPU framework on CIFAR-10 dataset. It has been taken around 2-h execution time for 50 epochs. However, 50 epochs are not enough to receive satisfied accuracy. The testing accuracy on different epochs has been listed in Table 5.3. For the sake of simplicity, we have shown the accuracy output for some sample epochs. The accuracy result is not very impressive due to the less number of epochs.

2. **Results on execution of Transfer learning on CIFAR-10 Dataset**

In this work, we have performed transfer learning on **ResNet50** and loaded pre-trained **ImageNet** weights to the model. We have retained all layers of **ResNet50** along with the final classification layers. The model is trained with

Table 5.3 Testing Accuracy ranges over different epoch samples in GAN

S. No.	Epoch samples	Testing accuracy (%)
1	5	53.26
2	10	54.33
3	15	55.25
4	20	55.10
5	25	54.38
6	30	56.53
7	35	56.98
8	40	58.12
9	45	59.71
10	50	60.29

Table 5.4 Training accuracy ranges over different epoch samples in transfer learning

S. No.	Epoch samples	Training accuracy (%)
1	1	91.76
2	2	94.02
3	3	94.50
4	4	95.09

a small number of epochs, and the results are listed in Table 5.4. It takes around 20 min to complete in the same colab framework.

Total params: 26,215,818

Trainable params: 26,162,698

Non-trainable params: 53,120.

The testing results for the validation dataset is shown in Fig. 5.14.

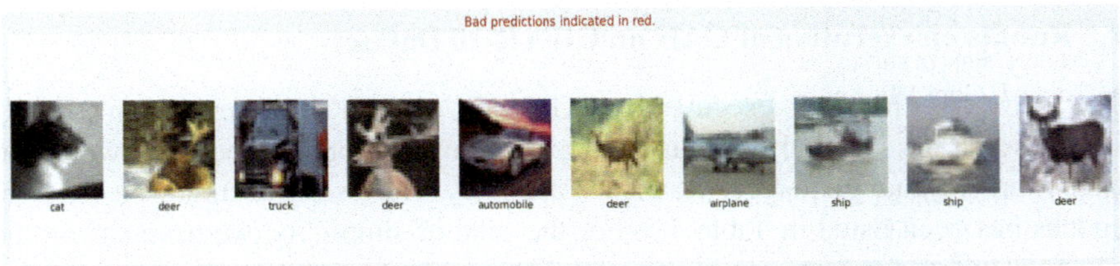

Fig. 5.14 Sample predictions on validation dataset

5.8 Conclusions

Multiclass and multi-label classification with deep learning models are explored in this chapter. A number of models that could be applied to this task, including transformers, CNN, VAE, LSTM, and GAN, are discussed. One method for classifying many labels is to convert each label into a binary form, meaning that it can either be present or absent. Another strategy is to split the multi-label problem into many binary classification problems using methods like classifier chains or binary relevance. Multi-label classification tasks can be accomplished with the help of models and tools from deep learning frameworks such as TensorFlow, PyTorch, and Keras. The pseudocode styles are mainly focused in this chapter. All things considered, deep learning models for multiclass and multi-label classification provide effective methods for resolving challenging classification issues when cases may concurrently belong to several classes or categories. This chapter shows how well deep learning models handle challenging classification tasks, yielding encouraging outcomes in multiclass and multi-label circumstances. The chapter emphasizes a number of methods, including MLP, GAN, recurrent neural networks, and convolutional neural networks, demonstrating how well they could capture complex correlations in the data. In addition, the chapter investigates methods such as LSTM, transfer learning, and hybrid models to enhance efficiency and tackle issues like high-dimensional feature spaces and class imbalance. All things considered, these results add to the expanding corpus of research on deep learning-based classification and provide insightful information for further study and application creation.

Bibliography

1. Read J, Perez-Cruz F (2014) Deep learning for multi-label classification. arXiv preprint arXiv: 1502.05988
2. Jabreel M, Moreno A (2019) A deep learning-based approach for multi-label emotion classification in tweets. Appl Sci 9(6):1123
3. Tarekegn AN, Ullah M, Cheikh FA (2024) Deep learning for multi-label learning: a comprehensive survey. arXiv preprint arXiv:2401.16549
4. Raju M, Gopi VP, Anitha VS (2021) Multi-class classification of Alzheimer's Disease using 3DCNN features and multilayer perceptron. In: 2021 Sixth international conference on wireless communications, signal processing and networking (WiSPNET). IEEE, pp 368–373
5. Hameed N, Shabut AM, Ghosh MK, Hossain MA (2020) Multi-class multi-level classification algorithm for skin lesions classification using machine learning techniques. Expert Syst Appl 141:112961
6. Kopinski T, Gepperth A, Handmann U (2015) A simple technique for improving multi-class classification with neural networks. In: ESANN
7. Balamurugan SP (2022) A comprehensive study on MLP and CNN, and the Implementation of multi-class image classification using deep CNN. In: machine learning and deep learning techniques for medical science. CRC Press, Boca Raton, pp 1–25
8. Heenaye-Mamode Khan M, Boodoo-Jahangeer N, Dullull W, Nathire S, Gao X, Sinha GR, Nagwanshi KK (2021) Multi-class classification of breast cancer abnormalities using deep convolutional neural network (CNN). PLoS ONE 16(8):e0256500

9. Mehmood A, Maqsood M, Bashir M, Shuyuan Y (2020) A deep Siamese convolution neural network for multi-class classification of Alzheimer disease. Brain Sci 10(2):84
10. Chaturvedi SS, Tembhurne JV, Diwan T (2020) A multi-class skin cancer classification using deep convolutional neural networks. Multimedia Tools Appl 79(39):28477–28498
11. Yu Y, Si X, Hu C, Zhang J (2019) A review of recurrent neural networks: LSTM cells and network architectures. Neural Comput 31(7):1235–1270
12. Lui HW, Chow KL (2018) Multiclass classification of myocardial infarction with convolutional and recurrent neural networks for portable ECG devices. Inform Med Unlocked 13:26–33
13. Schröder C, Niekler A (2020) A survey of active learning for text classification using deep neural networks. arXiv preprint arXiv:2008.07267
14. Khan MHM, Sahib-Kaudeer NG, Dayalen M, Mahomedaly F, Sinha GR, Nagwanshi KK, Taylor A (2022) Multi-class skin problem classification using deep generative adversarial network (DGAN). In: Computational intelligence and neuroscience
15. Zhang W, Gou Y, Jiang Y, Zhang Y (2022) Adversarial VAE with normalizing flows for multi-dimensional classification. In: Chinese conference on pattern recognition and computer vision (PRCV). Springer, Cham, pp 205–219
16. He H, Zhang J, Chen H, Chen X, Li Z, Chen X et al (2024) A diffusion-based framework for multi-class anomaly detection. In: Proceedings of the AAAI conference on artificial intelligence, vol 38, No. 8, pp 8472–8480
17. Kang Z, Yang B, Li Z, Wang P (2019) OTLAMC: an online transfer learning algorithm for multi-class classification. Knowl-Based Syst 176:133–146
18. Solomin AA, Ivanova YA (2020) Modern approaches to multiclass intent classification based on pre-trained transformers. J Sci Tech Inform Technol Mech Opt 128(4):532–538
19. Kumar R, Karthik K, Kamath SS (20) GAN-Based Encoder-Decoder Model for Multi-Label Diagnostic Scan Classification and Automated Radiology Report Generation. In: Handbook of AI-based models in healthcare and medicine. CRC Press, pp 93–109
20. Zhang Y (2018) Deep generative model for multi-class imbalanced learning
21. Ge Z, Demyanov S, Chen Z, Garnavi R (2017) Generative openmax for multi-class open set classification. arXiv preprint arXiv:1707.07418.
22. Ali-Gombe A, Elyan E (2019) MFC-GAN: Class-imbalanced dataset classification using multiple fake class generative adversarial network. Neurocomputing 361:212–221
23. Chu HM, Yeh CK, Wang YCF (2018) Deep generative models for weakly-supervised multi-label classification. In: Proceedings of the european conference on computer vision (ECCV), pp 400–415.
24. Gupta A, Narayan S, Khan S, Khan FS, Shao L, van de Weijer J (2023) Generative multi-label zero-shot learning. IEEE Trans Pattern Anal Mach Intell
25. Motamed S, Khalvati F (2021) Multi-class generative adversarial networks: improving one-class classification of pneumonia using limited labeled data. In: 2021 43rd Annual international conference of the IEEE engineering in medicine & biology society (EMBC). IEEE, pp 3817–3822
26. Shi W, Sheng VS, Li X, Gu B (2020) Semi-supervised multi-label learning from crowds via deep sequential generative model. In: Proceedings of the 26th ACM SIGKDD International conference on knowledge discovery & data mining, pp 1141–1149
27. Kim M (2015) Multiple-concept feature generative models for multi-label image classification. Comput Vis Image Underst 136:69–78
28. Mayee MK, Khanam MH (2024) Utilizing advanced deep learning for accurate multilabel classification of ocular diseases. In: 2024 1st International conference on cognitive, green and ubiquitous computing (IC-CGU). IEEE, pp 1–6
29. Singh I, Ningthoujam C, Verma VP, Roy B, Singh CT (2024) Transforming Multiclass to multilabel: advanced approaches in image classification. In: 2024 14th International conference on cloud computing, data science & engineering (confluence). IEEE, pp 458–463

Chapter 6
Applications of Multi-objective, Multi-label, and Multi-class Classifications

6.1 Introduction

Multi-label, multi-objective, and multi-class classification are essential fields for developing complicated decision-making procedures in a variety of applications. Optimizing numerous objectives at the same time is known as multi-objective classification, and it is crucial in situations where trade-offs between competing goals need to be managed, such as environmental management and health care. In text categorization, imagine annotation, and bioinformatics, multi-label classification—in which an instance might belong to more than one category—finds applications that allow for more detailed and nuanced predictions. In multi-class classification, examples are classified into many classes; this is fundamental to tasks like consumer segmentation, email classification, medical diagnosis, and handwriting and pattern recognition. When combined, these classification methods offer more detailed and adaptable predictive models that better suit a variety of challenging real-world scenarios, increasing their efficacy and relevance. Numerous real-world sectors find varied uses for multi-label classification. It helps in the diagnosis of several diseases from a single patient record in the healthcare industry. By grouping fraudulent transactions into several risk categories, it assists the finance industry in recognizing fraudulent transactions. By identifying films or music with numerous genres, entertainment services, such as Netflix, Amazon, and Spotify, use it to suggest content. It is used in social media to categorize comments or postings that cover a variety of themes or emotions. Furthermore, for tasks like text categorization in natural language processing—where documents are frequently linked to multiple subjects or tags—multi-label classification is essential. In many real-world applications, multi-objective classification—where decisions must balance many objectives—is essential. It allows for individualized treatment plans in the healthcare industry by taking medical expenses and patient results into account at the same time. Multi-objective classification helps portfolio management in finance by balancing the objectives of profit and risk. It streamlines production procedures in manufacturing to

S. Chakraborty and L. Dey, *Multi-objective, Multi-class and Multi-label Data Classification with Class Imbalance*, Springer Tracts in Nature-Inspired Computing,
https://doi.org/10.1007/978-981-97-9622-9_6

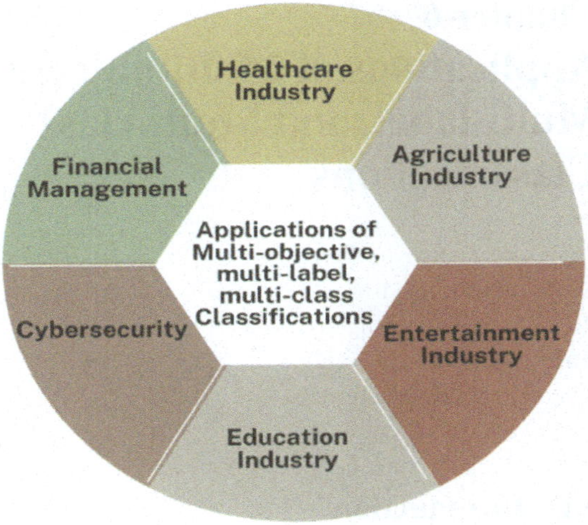

Fig. 6.1 Applications of classifications in various industry

raise product quality and reduce expenses. It is used in environmental management to achieve a balance between economic development and conservation activities. Multi-objective classification also aids in customer relationship management by helping to target marketing initiatives while taking maximizing revenue and customer satisfaction into account. This methodology guarantees comprehensive decision-making by simultaneously addressing several objectives.

In this chapter, we explore significant application areas where multiclass, multi-label, and multi-objective classifications are notably impactful. Figure 6.1 illustrates a range of compelling applications of these three classification types across various industries.

6.2 Applications in Healthcare Industry

In the healthcare sector, multi-class classification plays a key role in many applications by offering accurate diagnosis and categorization based on intricate medical data. This improves diagnostic precision while also supporting risk assessment, tailored therapy, and effective healthcare administration. Multi-class classification, which makes use of cutting-edge machine learning techniques, converts enormous volumes of medical data into useful insights that greatly enhance patient outcomes and operational effectiveness in the healthcare sector [1].

6.2.1 Diagnosis and Categorization of Diseases

1. Classification of cancers using histopathology images

Categorize various cancer kinds using histopathology images. For instance, identify the differences between invasive ductal carcinoma and lobular carcinoma, two kinds of breast cancer [2]. Employ dermoscopic images to categorize skin lesions into groups, such as benign, malignant, and different forms of skin cancer. In a research study [3], the diagnosis of breast cancer in the BreakHis dataset is studied using transfer learning techniques using two deep convolutional neural networks, namely Inception_V3 and Inception_ResNet_V2. Rotation, inversion, and various other data augmentation techniques are used to enlarge the BreakHis dataset in order to address the unequal distribution of samples of histological images of breast cancer. Because the Inception_ResNet_V2 network performed better on the original BreakHis dataset than the Inception_V3 network, it is selected to perform binary and multi-class classification diagnosis on the enlarged set of histological breast cancer images. For the linear kernel support vector machine (SVM) and one-nearest neighbor classifiers, features from the histopathology pictures of breast cancer are extracted using the Inception_ResNet_V2 network's potent feature extraction capacity. Inception_ResNet_V2 network-extracted image features are also fed into the K-means method for clustering analysis on the BreakHis dataset. Additionally, a novel autoencoder deep learning model is built to perform a K-means clustering analysis for the BreakHis dataset and employ a nonlinear transformation to the image features obtained by the Inception_ResNet_V2 network to extract the low-dimensional features of the image [3].

The two most common kinds of lung cancer are adenocarcinoma (LUAD) and squamous cell carcinoma (LUSC), and a skilled pathologist must visually check each to distinguish between the two. In a study [4], whole-slide images from The Cancer Genome Atlas are fed into a deep convolutional neural network (inception v3), which reliably and automatically classifies the images as either normal lung tissue, LUAD, or LUSC. Several distinct datasets, including biopsies and surgical resection specimens generated as frozen sections, are used to evaluate the performance of the suggested classification algorithms. In addition, the network is trained to anticipate the 10 LUAD genes with the most frequent mutations. Pathology images can be used to predict six of them: STK11, EGFR, FAT1, SETBP1, KRAS, and TP53. Figure 6.2 provides the entire training and testing procedures of the explained model [4].

2. Utilizing MRI or CT scan data

To distinguish between different neurological disorders, such as multiple sclerosis, multiple sclerosis, and ischemic brain stroke, Parkinson's disease [5], and Alzheimer's [6], we take the help of physiological signals and images [7]. This is a complex but critical task in medical diagnostics. Sophisticated categorization techniques, which frequently include deep learning and machine learning models, improve the precision and effectiveness of identifying these illnesses. These models examine a variety of datasets, such as biomarkers, clinical symptoms, genetic data,

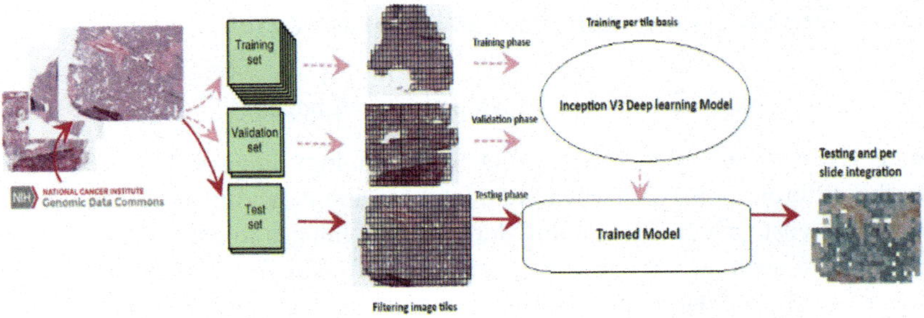

Fig. 6.2 Overall training and testing phases

and results from magnetic resonance imaging (MRI) and positron emission tomography (PET) scans [8]. For example, lesioning of the brain and spinal cord characterizes MS, degeneration of dopamine-producing neurons in the substantia nigra characterizes Parkinson's disease (PD), and brain amyloid plaques and neurofibrillary tangles characterize Alzheimer's disease (AD) [9]. These models may distinguish between various conditions by extracting and combining information from imaging data and clinical records. This allows for an accurate and timely diagnosis as well as the development of tailored treatment plans that enhance patient outcomes. For a thorough overview of the sources of data, types of data, deep learning models, and associated results, an intriguing review work [5] thoroughly summarizes published studies that applied advanced deep learning (DL) models to the diagnosis of Parkinson's disease (PD)-related symptoms and estimation of disease severity levels. In order to improve early clinical diagnosis based on motor symptoms, a multi-modal deep learning strategy is suggested in this systematic review. Patient outcomes can be enhanced, and early clinical identification of Parkinson's disease (PD) facilitated by the implementation of an accurate decision-making system based on DL models. In this work, a number of research have focused on gait, upper limb movement, and voice signals using both conventional and sophisticated methodologies in an effort to improve the accuracy of the model and diagnose Parkinson's disease.

A deep convolution neural network (DCNN) pipeline for the diagnosis of Alzheimer's disease and its phases utilizing MRI data scans is proposed in a paper [6]. Patients with Alzheimer's disease experience deterioration to their memory and cognitive abilities. Owing to the identical brain patterns and pixel intensity, this condition is very challenging to diagnose. A four-way classifier is introduced in this study to categorize individuals with AD, mild cognitive impairment (MCI), late mild cognitive impairment (LMCI), and healthy individuals. GoogleNet, ResNet18, and ResNet152 architectures are used for the classification of MRI images in this work. Figure 6.3 explains the overall process on MRI images.

By examining brain MRI data, a novel and quicker approach for Alzheimer's disease identification is proposed [8]. Four major stages (Non-demented, very mild AD, mild AD, moderate AD) of Alzheimer's disease is classified by using deep image convolution neural network in this work. This deep CNN network is tuned efficiently

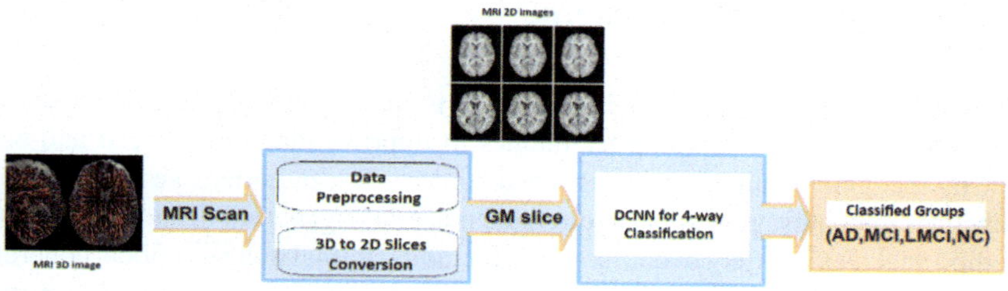

Fig. 6.3. 4-way classification of Alzheimer's disease using DCNN

by various hyperparameters and inspired by Inception-V4 network. In this network, the input MRI image is passed through a stem layer, which composed of a pair of convolution layers and max pooling layer. It is followed by a series of Inception layer stages, which consists of a number of filter-expansion layers, convolution layers, and average pooling layer. There are two reduction stages embedded among these three Inception stages. Each reduction stage consists of a number of filter-expansion layers, convolution layers, and max pooling layer. All of these modules' input and output go via a filter concatenation process. The last softmax layer has been altered for the purpose of classifying and detecting Alzheimer's disease in four different groups. The MRI images of Open Access Series of Imaging Studies (OASIS) dataset has been classified by CNN for detection of early-stage AD with n four different dementia stages [9]. Five convolutional blocks covered in feature engineering, max pooling, convolution layers, and classification are used in the CNN model. This work has employed Adam as the optimizer and cross-entropy as the loss function. Although softmax is connected to a mutually exclusive relationship, it has been utilized to categorize the multiclass AD stages. More than 80% of the tests are completed with accuracy. A small subset of datasets are provided when using computational techniques for diagnosis. As a result, our model kept the train, test, and validation datasets' image selection arbitrary. Three disease levels—normal, mild, and severe—are used to classify patients in another popular multiclass classification method [10] for the detection of AD. "Analyze Direct" and "ITK Snap" software are used to compute the volumes of the anatomical parts of the brain. Then, more features (age, gender, MMSE) are added and input into the nearest neighbor (KNN), decision tree, gradient boost, and random forest algorithms. The OASIS and ADNI datasets are used to train and test this methodology. Data augmentation, feature fusion, and feature extraction are also applied to preprocess the input datasets. With the OASIS dataset, random forest yields the maximum accuracy of 99% for white matter volume; when all three hippocampal volumes are combined, white matter and gray matter yield 98% accuracy. We are obtaining 92% accuracy for the gradient boost classifier for the ADNI dataset utilizing white matter volume, and we are also obtaining 92% accuracy after fusing all three volumes. For both databases, gradient boost has an accuracy of about 91%.

6.2.2 Retinal Medical Imaging

In ophthalmology, multiclass classification is an essential application that uses sophisticated machine learning techniques to improve diagnostic speed and accuracy by recognizing and classifying eye disorders. To categorize eye disorders such diabetic retinopathy, glaucoma, age-related macular degeneration, and cataracts, this method analyzes a variety of data sources, including retinal images, optical coherence tomography (OCT) scans, and clinical patient data. These systems are able to automatically and precisely identify between various eye disorders by training models on labeled datasets that contain parameters like macular thickness, optic nerve head morphology, and retinal vascular patterns. In the end, this improves patient outcomes and lessens the strain on healthcare systems by supporting individualized treatment programs and facilitating early identification and action. Using two popular techniques region growing and level set, retinal vessel segmentation has been done in this approach [11]. To enhance the vessels, the retinal vessel images undergo some preprocessing using the methods namely 2D Gabor wavelet and contrast-limited adaptive histogram equalization (CLAHE). Both the methods mentioned before are applied to extract retinal vessels. Using all three measurements, the performance is checked on the publicly available STARE and DRIVE databases where the observed results are as follows: accuracy of 95%, respectively. A research study introduces the deep convolutional neural network (DCNN) model for the detection of numerous diabetic eye diseases (DEDs) from retinal fundus images [12, 13]. It is able to discriminate between images of health and disease pathology. Data from open sources, such as the publicly accessible DRISHTI-GS, Messidor, Messidor-2, and Kaggle cataract datasets, is gathered. In the initial stage, mathematical morphology has been applied to provide contrast enhancement in the retinal fundus images. This work classifies five classes of DED, such as healthy, diabetic retinopathy (DR), diabetic macular edema (DME), glaucoma (Gl), and cataract (Ca) retinal fundus images (multi-class classification). The greatest accuracy of the suggested multi-class classification model is 81.33%. Developing an automated fundus image classification model that can identify a broader spectrum of ophthalmological illnesses is the goal of an intriguing research project [14]. The objective of this work is to classify ophthalmological illnesses using fundus images for multi-class multi-label classification. The right eye has anomalies associated with diabetic retinopathy and myopia, whereas the left eye appears to be normal. In order to address this challenging multi-class, multi-label classification challenge, our work [14] integrates information from fundus images of the patient's two eyes. It combines deep learning approaches for classification with their great performance and generalization capabilities. Two pre-trained convolutional neural networks (CNNs) based on transfer learning are introduced in order to classify fundus images from ODIR database into eight groups related to ophthalmological illnesses. For the purpose of multi-label multiclass classification of retinal images using Optical Coherence Tomography (OCT) data, a deep neural network is introduced in a research work [15]. The last four classifications include neovascular age-related macular degeneration, dry age-related macular degeneration, diabetic macular edema, and epiretinal membrane. In addition to enhancing multiclass, multi-label classification, this study adds factors, such as gender, age, and visual acuity to the pixel data. Comparing transfer learning to learning from a randomly started network is one of the study's goals. On previously taught Inception ResNet V2, which is trained using ImageNet data, transfer learning experiments are conducted. The final linear layer is trained to identify zero or more of the four disease groups, and pre-trained weights are loaded and frozen for the convolutional layers. The modified Inception ResNet V2 includes deleting the last softmax layer and modifying model settings so that it trains on grayscale images rather than RGB images. When allocating labels, the sigmoid takes the place of the softmax. RMSprop serves as both the MultiLabel Soft Margin Loss function and the optimizer. A few research studies [16] use pre-trained models, such as VGG16, VGG19, ResNet50, ResNet152, and DenseNet121 to conduct multiclass classification of eye disorders. With an amazing 95% accuracy rate, VGG19 is the best-performing model out of all the ones put to the test. In this work, 973 retinal images are used to test these models. They use applied data preprocessing with GAN model and transfer learning technique to generate more diverse ocular images for training. These programs are initially trained on a vast number of ImageNet ocular images, and they are responsible for recognizing four kinds of eye disorders: normal, diabetic, cataract-, and glaucoma-afflicted. The accuracy of this model is 89%. The segmented blood vessels must have their vital informative features extracted in order to correctly identify the images. A study [17] uses a hybrid classifier that combines the SqueezeNet and Long-Term Recurrent Convolutional Network (LRCN) to propose a novel multi-classification model for eye disorders (normal, retinitis pigmentosa, pathological myopia, maculopathy, and glaucoma). It segments blood vessels as a preprocessing step and then extracts characteristics from the input image. The model's performance is increased by using a normalizing methodology and applying a gradient-based contrast enhancement method to the retinal images. This unique model (FASSD-Net) is utilized for segmenting thin and thick lines of retinal vessels in order to overcome the inadequate segmentation efficiency when it comes to the crossover of blood vessels. Using a hybrid approach, eye problems are identified from retinal images.

Multiclass classification classifies various cell and tissue types through tissue sample analysis, which helps with disease diagnosis, such as cancer. This method involves analyzing histopathology images from tissue biopsies using machine learning algorithms. The system may learn to reliably categorize new samples by training these models on labeled datasets that comprise images of diverse cell and tissue types, such as benign tumors, malignant cancer cells, and normal cells. To distinguish between categories, characteristics such cell size, shape, texture, and spatial arrangement are retrieved. Pathologists can diagnose tumors more quickly and accurately thanks to this computerized classification, which also helps them determine the exact type and grade of cancer, allowing for more targeted and efficient treatment strategies. These algorithms' increased diagnostic accuracy improves patient outcomes and makes better use of available medical resources.

6.2.3 Personalized Treatment Plans

Personalized medicine is being revolutionized through multiclass classification using DNA profile-based patient classification through machine learning. This approach allows for exact therapy and dosage selection while lowering the risk of adverse medication reactions. Through the examination of genetic data, including single nucleotide polymorphisms (SNPs), gene expression levels, and enzyme activity, machine learning models are able to divide patients into different groups according to how they are expected to react to different drugs. These models can accurately anticipate how different individuals will react to particular therapies because they are trained on large datasets that link genetic variants to drug efficacy and side effects. This strategy guarantees that every patient receives the best medication and dosage based on their individual genetic profile, improving therapeutic efficacy, decreasing the need for prescription trial-and-error, and dramatically lowering the rate of adverse drug responses. Patient outcomes are enhanced as a result, and health care becomes more effective and individualized.

6.2.4 Mental Health

Machine learning and deep learning are used in multiclass classification, which uses professional interviews and patient questionnaires to reliably identify mood disorders such major depressive disorder (MDD), bipolar disorder, and anxiety disorders. Through the integration of extensive data from standardized patient questionnaires and qualitative insights from clinical interviews, these models are able to extract and analyze characteristics particular to each condition, including patterning of symptoms, severity, duration, and behavioral markers. These datasets are used to train deep learning architectures like convolutional neural networks (CNNs) and recurrent neural networks (RNNs), as well as machine learning methods like support vector machines (SVMs) and random forests, to discover complex patterns and correlations. Specifically, deep learning models are capable of processing big, intricate datasets and identifying subtle links within them. With the help of this advanced analysis, mood disorders can be precisely classified, enabling early and accurate diagnosis, individualized treatment planning, and ultimately improved patient outcomes in the field of mental health care.

A variety of cutting-edge machine learning techniques, including 1D CNN, 3D CNN, LSTM (long short-term memory), GRU (gated recurrent unit), and combinations like LSTM + GRU + RNN (recurrent neural network), are employed in a study [18] to build a multi-class stress identification model. The SWELL-KW dataset is used to train and test these models. The study investigates heart rate variability (HRV) features as putative stress biomarkers using this dataset. The setup of these principle components (PCA) in this work facilitates the selection and extraction of features, hence improving the effectiveness and efficiency of classification tasks.

LSTM is the most accurate neural network architecture for stress detection based on HRV features; it performed better than the other models. The study clarifies the complex relationship between ongoing stress and its possible physiological effects, highlighting the vital necessity of precisely identifying stress in order to manage mental health effectively. Although the study acknowledges that stress is a normal reaction to difficult circumstances, it also highlights the negative impacts of ongoing stress on mental health and how these effects can lead to common problems like anxiety, depression, and disturbed sleep.

A study [19] has been presented that uses text-based (English language data from Reddit) analysis to identify signs of mental health illnesses using a multi-class single-level (MCSL) classification algorithm inspired by the "divide and conquer" concept. It suggests a customized preprocessing method based on the dataset's properties to manage such unstructured data from Reddit. In this study [19], the MCML (multi-class multi-level) classification method is used to conduct binary classification at level 1 while taking into account two categories: mental illness and non-mental illness. Level 2 classifications for texts suggesting mental diseases included bipolar disorder, PTSD, ADHD, anxiety, and sadness. Following the necessary text preprocessing on Reddit data (such as removing sensitive information, transforming emoticons and emojis, removing numbers, removing punctuations), feature engineering (Term Frequency-Inverse Document Frequency and fast-text) has been used to convert strings into numerical values. It applies a machine learning-based methodology using a number of techniques, such as SVM, regression, random forest, and Naïve Bayes. To be more specific, SVM with different kernel functions is used in this investigation. The construction of LSTM and GRU models is the method used to apply the deep learning methodology. Furthermore, the MCML classification is carried out using a number of pre-trained BERT (bidirectional encoder representations from transformers) models (DistilBERT, ALBERT, and RoBERTa). The RoBERTa model provides the best accuracy in this work.

The objective of a research work [20] is to use multi-modal neuroimaging analysis to find brain biomarkers for bipolar disorder (BP) and schizophrenia (SZ). Multi-class classification models are developed for SZ, BP, and healthy controls using information from structural magnetic resonance imaging (sMRI), diffusion tensor imaging (DTI), and resting-state functional magnetic resonance imaging (rs-fMRI). First, features from sMRI, DTI, and rs-fMRI data are extracted, which total 272, 200, and 1875. The models are constructed using leave-one-out cross-validation and a one-against-one strategy utilizing a support vector machine (SVM) with recursive feature elimination (RFE). The left supramarginal gyrus's reduced regional functional connectivity strength (rFCS) and thicker left cuneus cortex are found to be important biomarkers.

This study's main objective [21, 22] has been to evaluate how well several machine learning-based algorithms identify complex patterns seen in datasets associated with ASD, encompassing a broad spectrum of diagnostic categories. According to the results, the logistic regression approach shows excellent accuracy, with rates in the binary classification system of 94.3% for children and 99% for teenagers. According to reports, the support vector machine (SVM) outperforms all other systems in a classification task that is only intended for adults, with an accuracy rate of 98.5%.

In a study [23], two quadruple and one octal classification are carried out using a deep learning (DL) approach in an effort to contribute to the settlement of this issue through various classifications of autism diagnosis (ASD) by taking age and gender aspects into consideration. Participants with TD (Typical Development) and ASD had their brain structural MRI (sMRI) scans preprocessed in the system utilizing the Canny Edge Detection technology. Next, using data augmentation (DA) techniques, the dataset is expanded five times. The grid search optimization (GSO) algorithm is used to create the best convolutional neural network (CNN) models. The system is intended to use three CNN models. Three models have been developed: the 1st is the quadruple classification model that considers gender; the 2nd is the quadruple classification model that considers age; and the 3rd is the eightfold classification model that considers both gender and age (model 3). The GSO algorithm is used for each model to determine which hyperparameters, out of the limit values, are best.

Sentiment analysis may make extensive use of the wealth of relevant information found on social media platforms. The data is scraped from Twitter to anticipate the forms of depression mentioned in the tweets. Five forms of depression (major, psychotic, atypical, postpartum, and bipolar) are predicted using tweets from the Twitter database based on lexical classification in a research study [24]. Explainable AI highlights the portions of tweets that correspond to different types of depression in order to provide reasoning. For training and feature extraction, bidirectional encoder representations from transformers (BERT) are utilized. The model is trained using deep learning and machine learning techniques. The overall process is summarized in Fig. 6.4.

By dividing complex medical data into discrete divisions, multi-class classification facilitates accurate and effective decision-making in the healthcare industry. This improves diagnostic precision while also supporting risk assessment, tailored therapy, and effective healthcare administration.

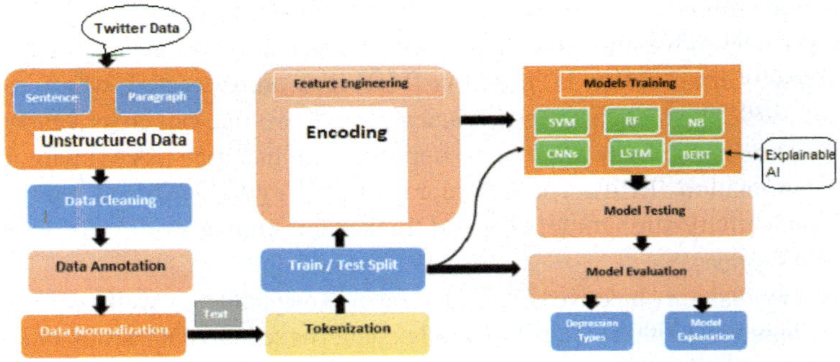

Fig. 6.4 Multiclass classification of depression from Tweets

6.3 Applications in Agriculture Industry

In the agricultural sector, multiclass multi-objective classification improves precision farming by concurrently optimizing several objectives across several classes. Applications include crop disease diagnostics, in which the system uses severity and treatment effectiveness to classify various diseases. Another use for the model is in agricultural yield prediction, where it classifies various yield levels and takes into consideration goals like resource efficiency and productivity maximization. Frequently used methods include multi-objective evolutionary algorithms and deep learning models, which are trained on some benchmark datasets for environmental data and plant health data. With the use of this strategy, farmers may make well-rounded decisions that balance a variety of objectives, including disease control [24–27], yield optimization [28–31], and sustainable resource use, which eventually results in more effective and sustainable agricultural practices. A brief discussion of classification tasks on different aspects of agriculture industry is given below.

6.3.1 Monitoring of Crop Health

Plant Leaf Disease Diagnosis: Plant disease diagnosis is made possible by the effective identification of several illnesses using a single model through the use of multi-class classification in machine learning (ML) and deep learning (DL). These methods can identify subtle patterns and symptoms associated with different plant diseases by training models on massive datasets of plant photos annotated with different disease labels. The capacity of convolutional neural networks (CNNs), a type of DL model, to automatically extract and learn features from images makes them particularly effective in this domain. This method improves early diagnosis and detection, enabling prompt intervention and treatment, and eventually enhancing crop health and output. Precision agriculture is aided by the integration of ML and DL in plant disease diagnostics, which offers scalable, dependable, and real-time monitoring systems. Three deep variants of CNN-SVM, ensemble algorithms, and densenet169 are presented in a research effort [25] that is appropriate for the identification and categorization of potato leaf diseases. This work [25] uses three and two layers of the deep ensemble model to classify only four types of potato leaf disease photos (early-blight disease, septoria disease, late-blight disease, and black-leg disease), along with one healthy leaf disease. This methodology consists of two stages: the training stage and the testing stage. Getting a crisp image of the potato leaf, labeling it, preprocessing it, and then training it with CNN-SVM and densener169 comprise the training step. The test image must then be gathered once more, preprocessed, and the disease must be identified and classified. Additionally, explainable AI models are used to explain the efficacy and reliability of the outcomes.

Plant disease identification presents both opportunities and obstacles, as highlighted by a study [26]. In accordance with this, it suggests an ensemble deep learning-based method for diagnosing plant diseases by combining deep learning-based models, such as VGG16, ResNet50, and AlexNet. By examining images of the plant leaves, it can accurately identify plant illnesses. To assess the robustness of the suggested method, a wide range of tests are carried out utilizing various plant leaf image datasets, including cherry, grape, maize, pepper, potato, strawberry, and cardamom. The suggested method achieved a maximum detection accuracy of 100% for binary datasets and 99.53% for multi-class datasets, according to experimental results.

A new deeper lightweight convolutional neural network architecture (DLMC-Net) is presented in a paper [27] to conduct plant leaf disease detection for real-time agricultural applications across multiple crops. To distinguish between the various kinds of leaf illnesses, a novel collective block and passage layer is included in this model. Using fewer parameters, the collective block extracts deeper features by utilizing the power of dense connectedness. To address the vanishing gradient issue, the output of the previous and current layers is combined and transmitted to the following layer in the passing layer. By lowering the learning rate, the vanishing gradient problem might occasionally diverge from the model training. The suggested DLMC-Net's passing layer normalizes the input often using batch normalization, making the model training process easier. This passage layer creates a narrower corridor for backpropagation by being positioned in between the two nearby collective blocks. Moreover, the ideal number of parameters is used in the design of the suggested DLMC-Net. Citrus, cucumber, grape, and tomato datasets that are accessible to the public are used for testing. Deeper features can be extracted without experiencing the vanishing gradient issue. The overall architecture of DLMC-Net is shown in Fig. 6.5.

In order to categorize and diagnose plant diseases, a work [32] aims to find a stacking ensemble-based solution by the application of many algorithms on image data. Employing a multi-layered perceptron as a meta classifier and the best four deep learning algorithms, the stacking ensemble is created. Following the required preprocessing and augmentation of the data, the data is divided into training and testing data, and then the model is constructed using a stacking ensemble approach. On the dataset, the classic deep learning architectures CNN, Xception, Inception V3, and ResNet50 all fared extremely well. As a result, the ensemble model accepts all four deep learning architectures as level 0 inputs. In the suggested model, a multi-layer perceptron is employed as a level 1 classifier to enhance the outcomes. The accuracy achieved by the suggested model is 98.13%. Because the suggested model is built on a stack-based ensemble method and starts with the 4 traditional models that execute the best, it outperforms standard models.

Identification of pests: By precisely identifying different plant pests from a single model using multiclass classification using machine learning (ML) and deep learning (DL) approaches, pest management strategies are improved. Through the use of large databases of plant photos labeled with various pest species, these technologies are able to precisely identify and categorize a variety of pests based on unique visual

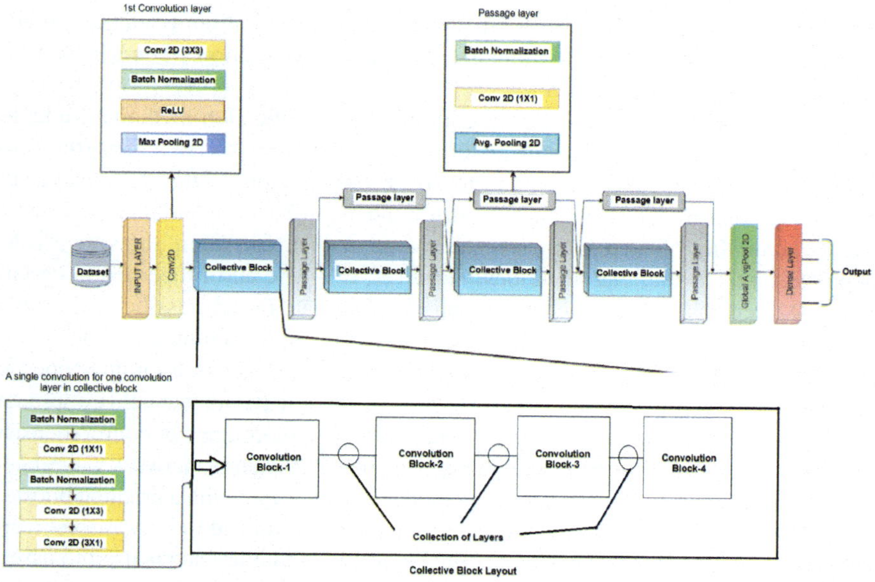

Fig. 6.5 DLMC-Net architecture

patterns. A subclass of DL models called convolutional neural networks (CNNs) perform particularly well in this challenge because of their ability to extract features and recognize patterns. This technique makes it easier to identify pests early and accurately, which enables prompt and focused actions that preserve plant health and maximize agricultural output. Using ML and DL together to identify pests promotes sustainable agriculture by giving agronomists and farmers access to reliable, scalable, and real-time monitoring tools [33].

6.3.2 Accurate Agriculture

Accurately classifying various soil types with multiclass classification through machine learning (ML) and deep learning (DL) approaches optimizes crop farming processes. These technologies are able to precisely detect and categorize soil based on its physical and chemical properties by training models on large datasets that comprise soil samples labeled with different categories, such as sandy, clay, loam, and silt. The capacity of convolutional neural networks (CNNs) and other deep learning models to extract intricate patterns and properties from soil images and sensor data makes them especially useful. By enhancing soil analysis, this method helps farmers better select crops and adjust farming activities to the unique soil types.

This improves crop output, soil health, and sustainable agricultural practices. Precision agriculture has a scalable, effective, and real-time solution when ML and DL are integrated for soil classification [28].

There are several machine learning-based soil classification methods available. Three primary factors—soil fertility, climate, and land revenue system—are the basis for the native classification that is employed, according to one of the research. Fertile and sterile soils are the two main categories under soil fertility classification. Based on the suitability of their crops and the source of their irrigation, they are further separated into several classifications. However, soil type and properties in certain climatic circumstances are the main emphasis of climate-based classification. Soil is classified primarily into three categories: dry, wet, and mild. A crop selection system [29] based on GBRT-based deep learning surrogate models has been developed to increase agricultural yields. Using information on soil parameters, this approach evaluates soil sites' suitability for particular crops in a thorough and comprehensive manner. The most ideal hyperparameters for the deep neural network have been found by combining a Bayesian optimization (BO) algorithm with a gradient boosted regression tree (GBRT). During the testing stage, the optimized hyperparameters are subsequently used. Furthermore, explainable artificial intelligence is used to assess how each input parameter affects each of the different outputs (XAI). Data collection and feature extraction are the first steps in the modeling process. Visualization of data and preprocessing come next. The model is then chosen using the preprocessed data after the hyperparameters have been changed. When used with statistical test analysis, the Pearson correlation coefficient helps to determine which features in the sample database are most valuable. The model is trained using the ideal qualities once they have been chosen. After that, a model built on Bayesian hyperparameter optimization incorporates the best hyperparameters for the GBRT-based DNN model.

A novel deep lightweight CNN model called Light-SoilNet network has been proposed in a research work [30]. By lowering the amount of learnable parameters, FLOPs, and minimizing overfitting in the process of classifying the soil images with improved performance, it can handle an unbalanced small soil image database. The classification of soil types linked to agricultural soils, such as sand, clay, loam, sandy loam, and loamy sand, is done in this work. To bring out the textural features in the soil sample photos, the database is subjected to HSV, RGB extraction, and adaptive histogram algorithms. A total of 392 soil samples from various agricultural fields are gathered to construct the soil image collection. The typical workflow consists of three main steps: preprocessing and augmentation, extraction of the ROI (region of interest) from the soil images, and classification of the soil images using Light-SoilNet. Thanks to its special convolution, the network is a nonlinear model with strong model characterization skills. The pooling structure may extract important features from complex input data. For two-dimensional soil images, 2D convolutional procedure is applied. The suggested model extracts the edges and corners of the soil images as its spatial features by using lower-size kernel filters. The fully connected, softmax, max pooling, dropout, ReLU, and batch normalization layers follow the convolution layer. Categorizing the soils with an overall accuracy of 97.2% has been achieved by the Light-SoilNet architecture.

6.3.3 Forecasted Yield

By forecasting and classifying crop yield levels, machine learning (ML) and deep learning (DL) algorithms optimize agricultural efficiency in multiclass yield levels resource allocation. These systems accurately divide fields into different production levels by analyzing a variety of information, such as soil quality, weather, crop health, and historical yield data. Recurrent neural networks (RNNs) and convolutional neural networks (CNNs) are examples of advanced models that can precisely allocate resources by recognizing intricate patterns and trends. This implies that distribution of inputs, such as pesticides, fertilizers, and water, can be done more efficiently if field regions are matched according to their expected yield potential. Thus, by making sure that resources are employed where they are most needed to obtain optimal agricultural yields, ML and DL increase productivity, decrease waste, and advance sustainable farming techniques [31].

6.3.4 Management of Fertilizers

By precisely categorizing crops based on their unique nutritional needs, artificial intelligence (AI), machine learning (ML), and deep learning (DL) can optimize fertilization procedures and maximize crop development while using less fertilizer. To precisely estimate the type and quantity of fertilizer required for each crop, sophisticated algorithms examine soil data, crop types, weather patterns, and growth stages. Different crops' nutrient uptake and growth responses are predicted by ML models, and these predictions can be further refined through the processing of complicated information by DL approaches. Farmers may apply fertilizers more effectively, cutting waste and environmental impact while improving crop quality and production overall, by putting this technique into practice. By maximizing resource utilization and supporting sustainable agriculture, this clever, data-driven strategy makes sure that every crop receives customized fertilizer management for optimum growth [34–36].

Artificial intelligence (AI)-based high-throughput phenotyping has been booming recently, and it can offer a useful way to identify fertilizer misuse. A novel multimodal fusion network (M2F-Net) is presented for high-throughput phenotyping to diagnose overload of fertilizers, driven by the power of deep learning paradigms and IoT sensors. Three techniques that integrate agrometeorological and image data are explored in a publication [34] by evaluating fusion at different phases. Initially, Multilayer Perceptron (MLP) on agrometeorological data and a pre-trained CNN model DenseNet121 on image data are used to construct two unimodal baseline networks. The multi-modal fusion network, which can proficiently learn from both image and non-image data, is constructed using these baselines. Prior to feeding the combined feature vector into any model, early fusion entails combining many modalities. A number of methods, including gated units, pooling, and concatenation, combine input modalities. The technique of incorporating learned, extracted characteristics from intermediate layers along with features from many modalities into a final layer is known as joint fusion. Joint fusion is often accomplished via neural networks. Integrating important predictions from several models to arrive at a final prediction is the fundamental concept behind late fusion. In this case, many modalities are used for individual models, and the aggregation function is used to integrate the predictions of multiple models to arrive at a final choice. We took the aggregation function to be average for the categorization. This fusion approach simultaneously acquires feature vectors from the non-image data and the crop image. In order to produce a final prediction, it also concatenates the obtained feature vectors from each modality.

With the use of an advanced machine learning paradigm, a work [35] attempts to examine the effects of excessive inorganic fertilizer on the growth, yield, and physiological characteristics of amaranthus. Agronomic data including moisture level, crop yield, plant height, leaf length, leaf breadth, and so on is measured and analyzed statistically that uses various NPK fertilizer dose regimens. This work seeks to assess the misuse of synthetic fertilizers by determining the productivity and adaptability of amaranthus in various chemically treated soil conditions and automatically phenotyping its features using image-based machine learning algorithms. With 80% accuracy, the SVM classifier is a better learning algorithm for the suggested system.

6.3.5 Management of Irrigation

Utilizing artificial intelligence (AI), machine learning (ML), and deep learning (DL) to manage irrigation transforms agricultural water use efficiency, guaranteeing ideal crop development while preserving water resources. To make accurate irrigation decisions, AI-driven systems examine large datasets, such as crop types, growth phases, soil moisture levels, and weather forecasts. While DL approaches can evaluate real-time sensor data to regulate irrigation in response to instant changes in environmental circumstances, machine learning (ML) models use previous data to forecast future water needs and optimize irrigation plans. With the help of these clever technologies, water can be delivered precisely and automatically to crops at precisely the appropriate times, cutting down on water loss and avoiding overwatering or underwatering. Farmers may increase crop yields, encourage sustainable water usage, and adjust to changing climatic conditions by utilizing these technology, which will result in more resilient and fruitful agricultural practices [37].

6.4 Applications in Entertainment Industry

In the entertainment sector, multiclass classification is essential for improving sentiment analysis, genre categorization, and content suggestion. These models can classify a variety of data by using techniques like decision trees, support vector machines

(SVMs), and convolutional neural networks (CNNs). They can also forecast movie genres, identify TV program themes, and classify user reviews and mood on social media. In studies, these models are trained using datasets, such as the MovieLens dataset for movie ratings and IMDb for full movie metadata. This strategy promotes engagement and happiness in a cutthroat market through increasingly individualized user experiences, focused marketing campaigns, and effective content management. Using AI, ML, and DL for multiclass multi-objective categorization in the entertainment sector improves several facets of content production, dissemination, and audience interaction. Artificial intelligence (AI) algorithms are able to classify content into various genres, target demographics, and popularity levels by evaluating massive volumes of data from numerous sources, including social media, streaming platforms, and user interaction logs. These models are able to optimize for numerous goals at once, such as improving content discoverability, raising subscription rates, and optimizing viewer pleasure. In order to offer material that suits individual tastes, machine learning algorithms, for example, can classify films and television series into genres (action, drama, comedy) and subgenres (romantic comedy, sci-fi thriller) based on user preferences and viewing history. Studios can choose future projects that strike a balance between creative merit and commercial potential with the use of DL techniques, especially neural networks, which can assess and forecast trends. Furthermore, audience input can be categorized into good, negative, or neutral sentiments utilizing sentiment analysis with NLP (natural language processing), offering insights into viewer reactions and directing marketing initiatives. The entertainment industry can enhance viewer satisfaction and participation by using multiclass multi-objective classification to understand content performance better, customize recommendations to different audience segments, and optimize content strategy to meet multiple business goals.

In a study [38], machine learning classifiers like logistic regression, stochastic gradient descent, decision trees, random forests, multinomial Naïve Bayes, and support vector machines are used to classify sentiments (sexual, religious, acceptable, and political) across multiple classes on Bangla social media comments. In order to eliminate data conflicts, data cleaning and preprocessing techniques are used in the initial phase to remove various noises from social media data. The natural language toolkit can be used to perform basic comment preprocessing, such as deleting numerals, punctuation, emojis, URLs, and usernames. This paper proposes the CNN-LSTM model combination, called CLSTM, in addition to the standard ML methods. While LSTM is specifically designed for sequential modeling, CNN may learn local responses from temporal or spatial data in the CLSTM model but not sequential correlations. This model's strength is its ability to use CNN to extract as many features as possible from texts and use those features as an LSTM input. Lastly, a web application named MBSA has been created to do multi-class classification on Bengali texts. It is based on the suggested CLSTM and the best baseline model, which is logistic regression. In the commercial sector, satisfaction detection is just as important as sentiment analysis using regional languages. A study [39] uses two well-known approaches, machine learning and deep learning (DL) techniques, to propose an application that uses Big Data from online businesses on social media, specifically Facebook and Twitter, to identify the Satisfaction tone that leads to customer pleasure. Following stemming, normalization, and stop word removal from the large data input, a variety of ML (KNN, SVM, NB, RF) and DL (CNN, RNN, and DNN) approaches are used for the classification task.

Multiclass and multi-label classification has a huge role in the sentiment analysis of movie reviews. Opinion mining, also known as sentiment analysis, is a technique used to obtain subjective data from movie reviews. This data includes the reviewer's suggestions, strengths and faults of the movie, and overall assessment of the movie. Potential viewers can use this information to make well-informed judgments about whether or not to watch a movie. A work [40] uses the Internet Movie Database (IMDB) Dataset and the Rotten Tomatoes dataset reviews to perform sentiment analysis of movie reviews using two pre-trained advanced language models: XLNet and bidirectional encoder representations from transformers (BERT). The advantages of autoencoder (AE) and generalized auto-regressive (AR) pre-training are combined in XLNet. The research [40] combines the capabilities of these sophisticated models with conventional ML techniques (logistic regression) to produce a comprehensive and potent tool for efficiently comprehending and classifying movie reviews.

In a study [41], several multi-class classification techniques are used to automatically assign ratings for customer movie reviews based on a 5-star rating system. The classification of reviews is done using five supervised algorithms (kNN, MNB, RF, XGBoost, and SVM) with tenfold cross-validation and two weighting methods (tf-idf and tf-igm) with an initial text preprocessing. The input dataset is prepared by gathering 10,000 movie reviews for various films from the Amazon website, using a 5-star rating system. Every rating group has two thousand reviews in it. A random selection of 1000 movie reviews is used as the test set. About 200 reviews make up each rating group in the test set. The specialists re-examine the evaluation results to confirm their accuracy. The remaining movie reviews are used as a training set once the test set has been obtained. The multi-class classifier model created by SVM and tf-igm produces the best results for automated ratings of customer reviews after confirming the efficacy of the employed techniques. A recurrent neural network (RNN) architecture called DSWE-GRNN, based on gated recurrent units (GRUs), is presented in a publication [42] for the multi-class review rating classification problem. As we typically do not have many reviews from the same person to gauge their leniency for a particular attitude, this model uses domain-specific word embeddings and is independent of the reviewer's information. A new framework for the rating-inference problem based on recurrent neural networks is introduced. Our model integrates domain-specific word embeddings for reducing dimensionality and the downsampling technique for data balance. The suggested model, which is independent of reviewers, will contribute to the improvement of review rating systems' intelligence by making precise star ratings recommendations based on historical data samples. The GRNN architecture is used because, when word embedding is used to train the model, it can extract contextual information from the provided text. In addition, the GRNN model uses less time than other recurrent neural network architectures, such as the LSTM. This model is evaluated using a number of baseline techniques, including CNN, LSTM, CNN-LSTM, and simple neural networks (SimpleNN) and on two popular benchmark datasets, i.e., IMDB dataset (with 10 classes) and Hotel Reviews dataset (with 5 classes). These days, it is hard to find qualitative information in a YouTube video. Based on the subscribers' comments, opinions, and personal experiences, a study [43] assists the subscribers in identifying the most experienced notion. Google APIs are used to gather the comments, which are then classified into many classes: Experience-positive, Experience-negative, Warning, Suggestions, Questions, and Praise. These courses assist us in applying machine learning techniques to find the qualitative analysis of the remarks. The suggested model demonstrates experimental results indicating the SVM, with an accuracy of 89.18%, is the best classifier. A work [44] develops a collection of machine learning (ML) techniques (SVM, NB) and deep learning-based models (LSTM, CNN) to identify multi-label sentiments and emotions for sentences containing Bangla and Romanized Bangla that are gathered from different YouTube videos. This work assigns a five-class (highly positive, positive, neutral, negative, and strongly negative) sentiment label to a Bangla sentence in addition to the three-class (positive, negative, and neutral) sentiment label. Additionally, it can identify any of the six fundamental emotions—anger, contempt, fear, joy, sadness, and surprise—from a Bangla sentence. However, the classification accuracy is not very attractive in this case. ML and DL classifiers are used in Web video categorization (WVC) that makes use of video textual information, visual features, or audio techniques [45]. It is based on user comments, YouTube information, and weighted attributes from video descriptions, just like the earlier studies. The required preprocessing and term weighting is completed before applying the classifiers (KNN, SVM, NB, RF, and DeepANN). Lastly, it conducts multiclass classifications using textual, audio, and visual types.

6.5 Applications in Education Sector

In the education sector, multiclass classification is crucial for improving academic performance analysis, automating administrative activities, and customizing learning experiences. Educational data can be classified into numerous classifications, such as forecasting student performance levels, classifying types of learning challenges, and classifying course materials, by using techniques like decision trees, support vector machines (SVMs), and neural networks. Experiments are conducted to train these models using datasets, such as the student performance dataset from the UCI Machine Learning Repository and large amounts of user data from educational platforms. This tool makes it possible to provide individualized instruction, identify students who require support early on, and allocate resources effectively, all of which contribute to the creation of a more productive and comfortable learning environment [46–52].

One of the most important performance indicators that teachers can use to keep an eye on their students' academic progress is their grade. To improve the accuracy of student grade prediction, addressing imbalanced datasets presents significant problems. In order to improve forecast accuracy, a publication [46] provides a thorough analysis of machine learning algorithms for predicting the final student grades in first semester courses. This research initially does a comparison analysis on real student data using six popular machine learning algorithms: LR, NB, J48, SVM, kNN, and RF. Then, to strengthen imbalanced multi-classification for student grade prediction, an amalgamation of two feature selection algorithms and an improvement on oversampling Synthetic Minority Oversampling Technique (SMOTE) is presented to automatically decide the sampling ratio with best picked features. In a published work [47], an unbalanced multi-classification approach based on feature integration and ensemble learning is proposed to forecast each college student's grant level. Prior to sorting the features by importance, the feature importance is prepared. Next, choose one of the six representative classification algorithms' ensemble rules to achieve better results. The outcomes of these individual classifiers are then combined using the majority vote ensemble rule. In order to find the best solutions and achieve a reasonable level of prediction accuracy, ensemble learning is used. Lastly, a comparison is made between the approach used in this work and six distinct classification algorithms: SVM, KNN, RF, AdaBoost (Ada), and GBDT. It has been discovered that having either too few or too many characteristics reduces classification accuracy. Moreover, feature importance selection can lead to very accurate classification. Ensemble learning can significantly increase classification accuracy when there are significant variances between the basis classifiers. Fusion using solely the powerful classifiers, Ada or GBDT, will experience overfitting because to insufficient training data. In high schools, a study [48] attempts to offer a method for choosing high-quality multiple-choice questions (MCQ) based on students' mid-semester test outcomes utilizing data mining techniques (SVM, Naïve Bayes, random forest, Decision Tree, Linear Regression, and KNN). The objective is to assess multiple-choice questions based on student profiles and topic midterm test outcomes using a data mining model with a difficulty index variable. This study builds on earlier research that developed the Cross-Industry Standard Process for Machine Learning (CRISP-ML) methodology, which aids in achieving the degree of interpretation required by stakeholders to produce effective solutions for the real world. We have tremendous opportunity to improve our understanding of idea production in asynchronous online discussions (AODs) with machine learning (ML) and deep learning (DL) techniques [51]. In a work [49], three DL systems and one baseline ML system are developed to automate message categorization while evaluating collaborative knowledge construction depth in academic AODs, using the interaction analysis model (IAM) as our theoretical foundation. For the first time, AOD messages from all IAM phases are categorized in one study. Four typical online courses were offered to show the viability of these platforms. Our accuracy score of 79% was the highest of all the scores obtained in all IAM phases. A conventional multiclass text categorization machine learning tool is the baseline system. The system vectorizes text using the TF-IDF technique. The random forest approach from a wide range of conventional machine learning classification algorithms is used. A BERT model that has been trained and adjusted to recognize word sequences in sentences makes up the final DL system. The BERT-base-uncased model, which has 109,482,241 total parameters, 12 transformer layers, and 768 dimensions, for pre-training is used. In BERT, a sequence can have a maximum length of 512 tokens. All AOD messages are only checked for the first 150 characters by our system, based on the average length of AOD messages from our corpus. Two dropout layers and three thick layers for fine-tuning are used. In deep learning models, the rectified linear unit is commonly used as an activation function in the first two dense layers, which consist of 64 and 32 neurons, respectively. The idea helps to explain how to encourage deeper contacts in AODs without inciting negative emotions in today's more divided and opinion-polarized society, given the complex and dynamic nature of collaboration. It provides symmetry in the ways that students seek out and give assistance as the cornerstone for involving them in critical thinking and enhancing their social skills. A work [50] uses machine learning approaches in order to create prediction models that can aid in the early identification of students who may drop out or not complete their degree on time. The major objective is to create a system that makes it feasible to identify students who are having academic difficulties or who are at risk of dropping out as soon as possible. A system like this will give academic institutions, administration, mentors, and instructors important information. The multiclass classification approach divides the target variables into three categories: the student who completed their degree on schedule, the enrolled student, and the dropped out student based on the first academic year data. After the data preprocessing along with handling imbalance dataset by SMOTE method, the random forest model is used to build the classification models (SMOTE + RF). In order to address the imbalanced nature of the data, two additional techniques are employed: the RusBoost algorithm (RB) and SVMSMOTE, which are followed by random forest (SVMSMOTE + RF). Furthermore, two algorithms that utilize the approaches for handling unbalanced datasets are applied. These are the Easy Ensemble classifier (EE) and the balanced random forest classifier (BRF). Five models are created for each of the datasets. A web-based career guidance system called EduKrishnaa [52] examines user profiles using technical and personality tests, suggests a variety of career paths, and offers educational resources on employment opportunities. It also shows a variety of skill-building projects, courses, internships, job openings, and startup opportunities. At present, the system helps recruiters, undergraduates, and students in the eighth and tenth grades. However, the goal is to expand its use to other fields and turn it into a one-stop shop for all career-related needs. The Howard Gardner Multiple Intelligence (Personality/Psychometric) Test and the Skill Test, which comprise technical questions from their respective disciplines, are the typical test procedures that are used to assess the users. For every one of the three student groups, different datasets are utilized. With eight variables, including scores for personalities, such as linguistic, logical, spatial, physical, musical, interpersonal, intrapersonal, and naturalist, the SSC student dataset comprises more than 3400 entries. The suggested task set is provided as an output by the target label. Similarly, the following datasets for the UG category are used to create machine learning models for Information Technology and Computer Science pre-final and final-year engineering students. Given that the dataset had more than two output classes, multiclass classification is used. Classifiers like KNN, RF, Naïve Bayes, SVM, and One-vs-Rest are employed. Random forest, one of the multiclass classification algorithms used by the system, yielded the highest accuracy of 95.45%.

6.6 Applications in Cybersecurity

In cybersecurity, multiclass categorization is essential for spotting and countering different kinds of online threats, which improves network security in general. Methods like decision trees, neural networks, and support vector machines (SVMs) are used to categorize various kinds of threats, including as phishing, malware, and denial-of-service (DoS) attacks. Research frequently makes use of datasets, such as the NSL-KDD and CICIDS2017, which offer extensive logs of network activity classified according to the type of attack. Cybersecurity systems can effectively identify and respond to a wide range of threats in real time by training models on these datasets. This lowers the risk of breaches and ensures the integrity of digital infrastructures. We discuss some interesting works on three key application areas in cybersecurity.

Key Applications

1. **Malware detection**: Features derived from file signatures, behavior patterns, and network activity are used by ML algorithms like decision trees, SVMs, and DL models like convolutional neural networks (CNNs) to categorize malware into groups (such as Trojans, worms, and spyware). Some suspicious activities and files that pass over the network may go undetected by intrusion prevention and detection systems. The eight most prevalent malware classes are classified using machine learning and deep learning models that have been developed [53]. A system's effectiveness in determining the type of malware, concentrating on the afflicted area, and implementing efficient countermeasures to reduce damages will increase with trained models that can classify malware classes. The recorded API call sequences can be analyzed by ML and DL models, which can then categorize the alleged program. In order to limit damages to victim computers, models that are trained to identify patterns of API sequence calls with the related malware class will ensure that the proper countermeasures are executed effectively and reduce the time it takes to identify the dangerous process. The goal of this research is to create multiclass classification models using a combination of deep learning and machine learning techniques to categorize eight key malware classes using sequential and tabular data. The data is initially preprocessed from the original dataset by removing a predetermined number of sequential API calls, normalizing the frequency count, and counting the frequency count. For categorization, standard machine learning methods like support vector machine, random forest, and Extreme Gradient Boosting are employed. In addition, deep learning methods including Transformer, one-directional CNN, MLP, and LSTM are employed. An ensemble of various neural networks is to be used in a study to identify malware [54]. This model seeks to precisely identify malware families by concatenating deep features gathered from malware images of three distinct models (the CNN architecture of VGGNet, AlexNet, and InceptionNet). The Least Absolute Shrinkage and Selection Operator (LASSO) method and

the support vector machine (SVM) classifier are employed as a feature selection technique to pick the minimal features required to improve the classification performance of an image-based malware detection system. Using only visual representations of the malware binaries, a unique machine learning approach is presented for malware detection and classification [55]. Principal component analysis is used to extract highly discriminative aspects of the malware category and structure in a confined subspace. Subsequently, the retrieved characteristics are classified into malware categories using an optimized support vector machine model. In contrast to current classification models, our approach uses straightforward algebraic dot products to categorize malware using representative digital images. This classifier is developed such that the extracted concise feature referring to two different malware categories should be significantly uncorrelated in order to effectively classify malware into distinct categories. However, this classifier should consider image features that are derived from many instances of the same malware class to be similar. Ultimately, improved classification performance results from combining the optimality of the suggested support vector machine (SVM) model with the discriminative power of the remaining image characteristics.

2. **Intrusion Detection Systems (IDS)**: Deep learning models, especially recurrent neural networks (RNNs) and long short-term memory networks (LSTMs) are used to identify and categorize anomalies suggestive of intrusion attempts, analyze real-time network traffic data. It has been demonstrated that intrusion detection systems based on deep learning are effective in spotting new threats. Convolutional neural networks (CNN) have not received much attention in research, despite the widespread usage of deep learning approaches for intrusion detection, such as stacked autoencoders (SAE) and deep belief networks (DBN). The effectiveness of CNN-based intrusion detection for recognizing various attack classes utilizing datasets like NSL-KDD and UNSW-NB 15 is assessed in a paper [56]. An essential component of maintaining Internet of Things (IoT) network security is the use of intrusion detection systems, or IDSs. However, creating a reliable, highly effective intrusion detection system (IDS) to identify particular irregularities in IoT devices and stop attacks is still a difficult undertaking. ANOVA, Kendall's test, mutual information, maximum relevance minimum redundancy, and chi-square are five filter-based feature selection techniques with different selection criteria that are used in a study to systematically analyze and compare two classifiers, XGBoost and LightGBM [57]. The Borda Count system and the Scott-Knott test are used to evaluate the NF-ToN-IoT-v2 dataset. For many applications, protecting network resources from unwanted activity is essential. In order to protect and identify network anomalies, a study [58] suggests a near real-time, multiclass, long short-term memory (LSTM)-based network intrusion detection system (NIDS) that makes use of complicated cloud CSE-CICIDSS2018 datasets. The suggested approach reduces dimensionality and selects features using a random forest method. Additionally, an LSTM model based on deep learning and designed to detect network intrusions uses the best attributes that are chosen.

3. **Phishing Attack Classification**: Malicious URLs are a common hazard to businesses, social networks, and online banking these days. By examining textual content, metadata, and link architectures, natural language processing (NLP) approaches paired with machine learning (ML) models identify emails and URLs as valid, phishing, or spam. Therefore, it becomes essential to identify the type of attack and implement a successful defense. A research based on multi-class classification suggests an approach to identify malicious URLs and the kind of attacks [59]. 42 fresh features of malware, phishing, and spam URLs are suggested in this work. The detection efficiency and attack-type identification of malicious URLs of common attack kinds, such as spamming, phishing, and malware distribution, are enhanced by these characteristics. Modern supervised batch and online machine learning classifiers are used in experiments on the binary and multi-class dataset. Three parts comprise the methodology: feature extraction, preprocessing and labeling, training using online and batch learning algorithms for multiple classes, and detecting attack types. The classification techniques utilized in this work include multi-class confidence weighted learning (MC-CW), OVA L1-reg L2-loss SVM (One-vs-all (OVA) SVM), and OVO L1-reg L2-loss SVM (One-vs-one (OVO) SVM). All these methods need us to preserve the URL features as sparse vectors. Using a binary setting, the maximum accuracy of 99.86% is attained in recognition of harmful URLs, and an average accuracy of 98.44% is attained in the detection of attack types using a confidence weighted learning classifier in a multiclass setting with the input URL attributes. The primary reason for assaults like drive-by downloads, XSS, malware delivery, etc., is the absence of analysis and detection of obfuscated JavaScripts in the webpages, which is an issue in this methodology. Given that short URLs are currently the most popular trend on microblogging platforms like Facebook, Twitter, and others, further research into their properties is necessary for identification of attack types and for efficient detection of attacks. Another study [60] uses URL-based features in a multiclass classification environment to identify fraudulent URLs. Three common forms of URL attacks—phishing, spam, and malware—have been focused. Without using external features from outside sources, every feature used here is taken straight from the URL string. It can be integrated into new or pre-existing anti-spam, anti-phishing, and anti-malware detection systems as an additional tool. For the multiclass classification, a set of ensemble learners is utilized, including Extreme Gradient Boosting (XGBoost), Adaptive Boosting (AdaBoost), Light Gradient Boosting (LightGBM), and Categorical Boosting (CatBoost). The key characteristics in this work are derived from word-based features, bag of words segmentation, and Kullback–Leibler Divergence (KL divergence). Both with and without the experimental features, the models are tested. When compared to other datasets, spam detection receives the lowest scores in every experiment.

Benefits

- **Enhanced Detection Accuracy**: Security systems can reduce false positives and improve overall security posture by classifying threats into distinct categories and providing more precise and customized responses.
- **Automated Reaction**: The utilization of multiclass categorization facilitates automated security reactions, which include the isolation of compromised systems, obstructing malevolent IP addresses, and warning administrators about particular categories of risks.
- **Threat Intelligence**: Thorough categorization offers important information about the characteristics and frequency of various attack kinds, which helps to guide proactive security steps and threat intelligence initiatives.

6.7 Applications in Financial Industry

In the financial sector, multiclass classification plays a critical role in improving risk management, credit scoring, and fraud detection. Support vector machines (SVMs), neural networks, and random forests are examples of sophisticated algorithms that are used to categorize investment risks, consumer profiles, and transactions into different groups. Datasets with a variety of examples of financial transactions and consumer behavior, such as the German Credit Dataset and the Credit Card Fraud Detection Dataset, are commonly used in experiments. Financial organizations may evaluate creditworthiness, detect fraudulent activity with greater accuracy, and better manage investment risks by training models on these datasets. This can result in enhanced decision-making, heightened security, and optimized financial services [61]. Associative classification (AC) multi-label rule generation from single-label datasets is regarded as a difficult task. The Enhanced Multi-label Classifiers-based Associative Classification (eMCAC) technique is presented in a study [62] and is able to infer rules related to a group of classes from single-label data that other existing AC algorithms are not able to. In addition, by employing a classifier building technique, eMCAC minimizes the quantity of extracted rules. The three primary steps of the algorithm are class assignment, classifier construction, and rule discovery. eMCAC searches and extracts rules from the training dataset iteratively in the first stage. Step two involves discarding redundant rules, resulting in a lack of training data coverage for the rules. The classifier, which has rules in it, is the result of the second stage. To determine the classifier's prediction rate, the test dataset is used in the final step of the process. According to the suggested algorithm, all potential values for each of the categorical input attributes in the training dataset are mapped to a collection of positive integers. Any discretization technique can be used for continuous attributes prior to the training stage. A real-world application dataset pertaining to online phishing has been used to test the suggested method. The assessment metrics are label-weight, any-label, accuracy, and number of rules. From the phishing data, the eMCAC algorithm can generate multi-label rules, with each training example belonging to a single class. Using the chi-square feature selection technique, a smaller useful feature set is found for identifying the type of website. Even if corporate financial audit controls have improved over time, there are still more cases of corporate fraud, which makes it necessary to look into the elements that could serve as early warning signs and create efficient systems for spotting financial fraud. Based on the fraud diamond theory, financial statements from 133 Greek firms listed on the Athens Stock Exchange between 2014 and 2019 are examined in a study [63]. This research assessed the performance of models built using standard machine learning techniques in a multi-label classification situation and suggested an improvement of fraud detection models based on the fraud diamond hypothesis. Both non-financial and frequently used financial variables form the basis of the analysis. Six different types of auditors' comments are examined in order to identify cases of falsified financial statements (FFS) in the sample. These comments are then sorted into groups using a hierarchical clustering approach, defining meaningful classes made up of various FFS types. The suggested multi-label categorization framework's labels and the generated classes match. To differentiate between FFS and nFFS situations and to recognize the various comments given by auditors in FFS cases, a set of binary classifiers (LR, KNN, RF, and generalized additive models) has been utilized. Furthermore, additional models (multi-label NN algorithm (NNML), multi-label-kNN (MLKNN), and MLKNN + NNML) are constructed in the case of multi-label classification in order to distinguish between cases that belong in each of the FFS categories and the other observations. While the MLKNN outperforms the LR base model, the two corrective methods together yield better results than LR, especially when the NNML technique is used. Banks and other financial institutions have raised their standards for classifying customers' credit risk levels in response to the rise in loan demand. This is done in an effort to improve loan decisions and amount allocation while lowering pre-loan risk. A Multi-level Classification-based Ensemble and Feature Extractor (MLCEFE) that combines the benefits of ensemble classification, feature extraction, and sampling is proposed in a research work [64]. After resolving the data imbalance issue with SMOTE + Tomek connections, MLCEFE employs principal component analysis (PCA), autoencoders (AE), and deep neural networks (DNNs) to convert the initial variables into higher-level abstract features for feature extraction. Ultimately, it integrated several ensemble learners to enhance the impact of multi-classification of personal credit risk. It offers suggestions for learning personal credit features in addition to demonstrating the deep model's potent learning capabilities. The combining of ensemble learning models is the subject of the next section. In this article, four popular ensemble learning models—random forest (RF), XGBoost, LightGBM, and GDBT—are taken and combined using stacking techniques. Finally, to enhance the classification effect of personal credit risk, the MLCEFE integrates feature extractors with the ideal ensemble model.

6.8 Conclusions

This chapter explored the noteworthy developments in the application of multi-label, multi-objective, and multi-class classifications in a range of industries, such as cybersecurity, health care, education, entertainment, agriculture, and finance. Because these advanced categorization techniques offer more accurate, nuanced, and extensive analytical capabilities, they have revolutionized every discipline. They improve threat detection and system protection in cybersecurity, and they improve diagnostic and therapy personalization in health care. They improve performance evaluation and personalized learning pathways in different stages of education, and they improve content recommendations in entertainment. These techniques improve production prediction and crop health monitoring in agriculture, and they support fraud detection and risk management in finance. The amalgamation of these sophisticated classification techniques underscores their adaptability and revolutionary capacity to tackle complex, practical challenges in various sectors.

Bibliography

1. Renuka P, Booba B (2022) Survey of machine learning algorithms for disease diagnosis multi-class classification of medical data. NeuroQuantology 20(11):4290
2. Demir C, Yener B (2005) Automated cancer diagnosis based on histopathological images: a systematic survey. Rensselaer Polytechnic Institute, Technical Report
3. Xie J, Liu R, Luttrell J IV, Zhang C (2019) Deep learning based analysis of histopathological images of breast cancer. Front Genet 10:80
4. Coudray N, Ocampo PS, Sakellaropoulos T, Narula N, Snuderl M, Fenyö D et al (2018) Classification and mutation prediction from non–small cell lung cancer histopathology images using deep learning. Nat Med 24(10):1559–1567
5. Skaramagkas V, Pentari A, Kefalopoulou Z, Tsiknakis M (2023) Multi-modal deep learning diagnosis of Parkinson's disease—a systematic review. IEEE Trans Neural Syst Rehabil Eng
6. Farooq A, Anwar S, Awais M, Rehman S (2017) A deep CNN based multi-class classification of Alzheimer's disease using MRI. In: 2017 IEEE International conference on imaging systems and techniques (IST). IEEE, pp 1–6
7. Raghavendra U, Acharya UR, Adeli H (2020) Artificial intelligence techniques for automated diagnosis of neurological disorders. Eur Neurol 82(1–3):41–64
8. Islam J, Zhang Y (2017) A novel deep learning based multi-class classification method for Alzheimer's disease detection using brain MRI data. In: Brain informatics: international conference, BI 2017, Beijing, China, 16–18 Nov 2017. Springer, Berlin, pp 213–222
9. Battineni G, Chintalapudi N, Amenta F, Traini E (2021) Deep learning type convolution neural network architecture for multiclass classification of Alzheimer's Disease. In: Bioimaging, pp 209–215
10. Biswas R, Gini JR (2024) Multi-class classification of Alzheimer's disease detection from 3D MRI image using ML techniques and its performance analysis. Multimedia Tools Appl 83(11):33527–33554
11. Zhao YQ, Wang XH, Wang XF, Shih FY (2014) Retinal vessels segmentation based on level set and region growing. Pattern Recogn 47(7):2437–2446
12. Sarki R, Ahmed K, Wang H, Zhang Y, Wang K (2021) Convolutional neural network for multi-class classification of diabetic eye disease. EAI Endorsed Trans Scalable Inform Syst 9(4)

13. Sarki R, Ahmed K, Wang H, Zhang Y (2020) Automated detection of mild and multi-class diabetic eye diseases using deep learning. Health Inform Sci Syst 8(1):32
14. Gour N, Khanna P (2021) Multi-class multi-label ophthalmological disease detection using transfer learning based convolutional neural network. Biomed Signal Process Control 66:102329
15. Mehta P, Lee A, Lee C, Balazinska M, Rokem A (2018) Multilabel multiclass classification of OCT images augmented with age, gender and visual acuity data. BioRxiv, 316349
16. Mayee MK, Khanam MH (2024) Utilizing advanced deep learning for accurate multilabel classification of ocular diseases. In: 2024 1st International conference on cognitive, green and ubiquitous computing (IC-CGU). IEEE, pp 1–6
17. Alharbi M (2024) Multi-classification of eye disease based on fundus images using hybrid squeeze net and LRCN model. Multimedia Tools Appl 1–30
18. Kumar CS, Bhaskaru O, Karanam B, Ahmed AA, Reddy YGCT, Sathwik M (2024) Exploring multi-class stress detection using deep neural networks. In: 2024 International Conference on Cognitive Robotics and Intelligent Systems (ICC-ROBINS). IEEE, pp 69–74
19. Sarno R, Ghozali I (2024) Multi-class multi-level classification of mental health disorders based on textual data from social media. J Inform Commun Technol 23(1):77–104
20. Chen M, Xia X, Kang Z, Li Z, Dai J, Wu J et al (2024) Distinguishing schizophrenia and bipolar disorder through a multiclass classification model based on multimodal neuroimaging data. J Psychiatric Res
21. Bawa P, Kadyan V, Mantri A, Vardhan H (2024) Investigating multiclass autism spectrum disorder classification using machine learning techniques. e-Prime-Adv Electr Eng Electron Energy 8:100602
22. Alves CL, Martinelli T, Sallum LF, Rodrigues FA, Toutain TGDO, Porto JAM et al (2024) Multiclass classification of autism spectrum disorder, attention deficit hyperactivity disorder, and typically developed individuals using fMRI functional connectivity analysis. medRxiv 2024-04.
23. Nogay HS, Adeli H (2024) Multiple classification of brain mri autism spectrum disorder by age and gender using deep learning. J Med Syst 48(1):15
24. Nusrat MO, Shahzad W, Jamal SA (2024) Multi class depression detection through tweets using artificial intelligence. arXiv preprint arXiv:2404.13104
25. Paul H, Ghatak S, Chakraborty S, Pandey SK, Dey L, Show D, Maity S (2024) A study and comparison of deep learning based potato leaf disease detection and classification techniques using explainable AI. Multimedia Tools Appl 83(14):42485–42518
26. Sunil CK, Jaidhar CD, Patil N (2022) Binary class and multi-class plant disease detection using ensemble deep learning-based approach. Int J Sustain Agric Manage Inform 8(4):385–407
27. Sharma V, Tripathi AK, Mittal H (2023) DLMC-Net: Deeper lightweight multi-class classification model for plant leaf disease detection. Eco Inform 75:102025
28. Azmin SNFLM, Arbaiy N (2022) Soil classification based on machine learning for crop suggestion. J Soft Comput Data Mining 3(2):79–91
29. Bhat SA, Hussain I, Huang NF (2023) Soil suitability classification for crop selection in precision agriculture using GBRT-based hybrid DNN surrogate models. Eco Inform 75:102109
30. Pandiri DK, Murugan R, Goel T (2024) Smart soil image classification system using lightweight convolutional neural network. Expert Syst Appl 238:122185
31. Mohyuddin G, Khan MA, Haseeb A, Mahpara S, Waseem M, Saleh AM (2024) Evaluation of machine learning approaches for precision farming in smart agriculture system—a comprehensive review. IEEE Access
32. Sharma A, Dalmia R, Saxena A, Mohana R (2024) A stacked deep learning approach for multiclass classification of plant diseases. Plant Soil 1–18
33. Mekha J, Parthasarathy V (2022) An automated pest identification and classification in crops using artificial intelligence—a state-of-art-review. Autom Control Comput Sci 56(3):283–290
34. Dhakshayani J, Surendiran B (2023) M2F-Net: A deep learning-based multimodal classification with high-throughput phenotyping for identification of overabundance of fertilizers. Agriculture 13(6):1238

35. Dhakshayani J, Jyothsna J, Surendiran B (2024) Fertilizer optimization through machine learning-driven models: an empirical investigation on smart farming of Amaranth. Int J Environ Climate Change 14(5):254–267
36. Tanaka TS, Heuvelink GB, Mieno T, Bullock DS (2024) Can machine learning models provide accurate fertilizer recommendations? Precis Agric 1–18
37. Umutoni L, Samadi V (2024) Application of machine learning approaches in supporting irrigation decision making: a review. Agric Water Manag 294:108710
38. Haque R, Islam N, Tasneem M, Das AK (2023) Multi-class sentiment classification on Bengali social media comments using machine learning. Int J Cogn Comput Eng 4:21–35
39. Kanan T, Mughaid A, Al-Shalabi R, Al-Ayyoub M, Elbes M, Sadaqa O (2023) Business intelligence using deep learning techniques for social media contents. Clust Comput 26(2):1285–1296
40. Danyal MM, Khan SS, Khan M, Ullah S, Mehmood F, Ali I (2024) Proposing sentiment analysis model based on BERT and XLNet for movie reviews. Multimedia Tools Appl 1–25
41. Polpinij J, Luaphol B (2021) Comparing of multi-class text classification methods for automatic ratings of consumer reviews. In: International conference on multi-disciplinary trends in artificial intelligence. Springer, Cham
42. Hassan J, Shoaib U (2020) Multi-class review rating classification using deep recurrent neural network. Neural Process Lett 51(1):1031–1048
43. Nawaz S, Rizwan M, Yasin S, Ahmed M, Farooq U (2020) Multi-class classification of the YouTube comments using machine learning. Pak J Eng Technol 3(2):183–188
44. Tripto NI, Ali ME (2018) Detecting multilabel sentiment and emotions from bangla youtube comments. In: 2018 international conference on Bangla speech and language processing (ICBSLP). IEEE, pp 1–6
45. Yafooz WM, Alsaeedi A, Alluhaibi R, Abdel-Hamid ME (2022) Enhancing multi-class web video categorization model using machine and deep learning approaches. Int J Electr Comput Eng 12:3176
46. Bujang SDA, Selamat A, Ibrahim R, Krejcar O, Herrera-Viedma E, Fujita H, Ghani NAM (2021) Multiclass prediction model for student grade prediction using machine learning. IEEE Access 9:95608–95621
47. Sun Y, Li Z, Li X, Zhang J (2021) Classifier selection and ensemble model for multi-class imbalance learning in education grants prediction. Appl Artif Intell 35(4):290–303
48. Sucipto S, Prasetya DD, Widiyaningtyas T (2024) Educational data mining: multiple choice question classification in vocational school. MATRIK: Jurnal Manajemen, Teknik Informatika dan Rekayasa Komputer 23(2):379–388
49. Eryilmaz E, Thoms B, Ahmed Z (2024) Theory-guided multiclass text classification in online academic discussions. J Comput Inform Syst 1–12
50. Martins MV, Baptista L, Machado J, Realinho V (2023) Multi-class phased prediction of academic performance and dropout in higher education. Appl Sci 13(8):4702
51. Vijayalakshmi V, Venkatachalapathy K (2019) Deep neural network for multi-class prediction of student performance in educational data. Int J Recent Technol Eng 8(2):5073–5081
52. Ajgaonkar S, Tale P, Joshi Y, Jore P, Jakate M, Lavangare S, Kadam D (2023) EduKrishnaa: a career guidance web application based on multi-intelligence using multiclass classification algorithm. In: International conference on multi-disciplinary trends in artificial intelligence. Springer, Cham, pp 601–610
53. Adair S, Varol C, Liang F, Pham V (2024) Malware API call-based multiclass-classification using machine learning and deep learning. In: 2024 12th International symposium on digital forensics and security (ISDFS). IEEE, pp 1–7
54. Haile MB, Walle YM, Belay AJ (2024) Enhanced Image-based malware multiclass classification method with the ensemble model and SVM. Open Inform Sci 8(1):20240003
55. Ghouti L, Imam M (2020) Malware classification using compact image features and multiclass support vector machines. IET Inf Secur 14(4):419–429
56. Potluri S, Ahmed S, Diedrich C (2018) Convolutional neural networks for multi-class intrusion detection system. In: Mining intelligence and knowledge exploration: 6th international

conference, MIKE 2018, Cluj-Napoca, Romania, 20–22 Dec 2018. Springer, Berlin, pp 225–238
57. Hamdouchi A, Idri A (2024) Multiclass intrusion detection in IoT using boosting and feature selection. In: World conference on information systems and technologies. Springer, Cham, pp 128–137
58. Vibhute AD, Khan M, Kanade A, Patil CH, Gaikwad SV, Patel KK, Saini JR (2024) An LSTM-based novel near-real-time multiclass network intrusion detection system for complex cloud environments. Concurr Comput Practice Exp 36(11):e8024
59. Patil DR, Patil JB (2018) Feature-based malicious URL and attack type detection using multi-class classification. ISeCure 10(2)
60. Manyumwa T, Chapita PF, Wu H, Ji S (2020) Towards fighting cybercrime: Malicious url attack type detection using multiclass classification. In: 2020 IEEE international conference on big data (Big Data). IEEE, pp 1813–1822
61. Pattnaik D, Ray S, Raman R (2024) Applications of artificial intelligence and machine learning in the financial services industry: a bibliometric review. Heliyon
62. Abdelhamid N (2015) Multi-label rules for phishing classification. Appl Comput Inform 11(1):29–46
63. Tragouda M, Doumpos M, Zopounidis C (2024) Identification of fraudulent financial statements through a multi-label classification approach. Intell Syst Account Finan Manage 31(2):e1564
64. Wang Y, Wu Z, Gao J, Liu C, Guo F (2024) A multi-level classification based ensemble and feature extractor for credit risk assessment. PeerJ Comput Sci 10:e1915

The manufacturer's authorised representative in the EU is Springer Nature Customer Service Centre GmbH, Europaplatz 3, 69115 Heidelberg, Germany. If you have any concerns regarding our products, please contact ProductSafety@springernature.com

Printed and bound by CPI Group (UK) Ltd, Croydon, CR0 4YY
01/10/2026
02225919-0011